Stratigraphy and Depositional History of the Star Peak Group (Triassic), Northwestern Nevada

K. M. Nichols
N. J. Silberling
Department of Geology, Stanford University
Stanford, California 94305
(Present address: U.S. Geological Survey
Denver, Colorado 80225)

THE
GEOLOGICAL SOCIETY
OF AMERICA

SPECIAL PAPER 178

Library of Congress Catalog Card Number 77-89753
ISBN 0-8137-2178-4

Published by
THE GEOLOGICAL SOCIETY OF AMERICA, INC.
3300 Penrose Place
Boulder, Colorado 80301

Printed in the United States of America

Contents

ILLUSTRATIONS

Acknowledgments

Studies of the depositional and tectonic history of lower Mesozoic rocks in the Great Basin have been supported by National Science Foundation Grants GA 1398 (1968-1970), GA 19424 (1970-1972), and GA 36105 (1973-1974) to Stanford University. Investigations by Nichols in 1969 were supported by a Texaco Graduate Fellowship to Stanford University.

We thank A. K. Armstrong, P. R. Hoover, J. H. Stewart, B. R. Wardlaw, and D. H. Whitebread for critically reading the manuscript.

Abstract

The Star Peak Group in and near the Sonoma Range one-degree quadrangle, northwestern Nevada, is redefined to include the mainly calcareous strata that overlie the predominantly volcanic rocks of the Koipato Group and in turn are overlain by the terrigenous clastic rocks of the Auld Lang Syne Group. The Star Peak Group ranges in age from late Spathian (latest Early Triassic) to late Karnian (middle Late Triassic), and it is as much as 1,200 m thick. Although the paleogeography of the group has been affected locally by post-Triassic low-angle faulting, widespread formational units within the Star Peak genetically relate all parts of its outcrop region and obviate the need to separate these rocks into different major structural blocks, as has been done in the past.

In the older parts of the Star Peak Group, complex stratigraphic patterns among a variety of carbonate and terrigenous clastic rocks resulted from localized relative uplift, first early and then late in Middle Triassic time. Widespread diagenetic secondary dolomitization of calcareous rocks in the older part of the section further complicates the primary stratigraphic patterns. Following the earlier of these intra-Star Peak tectonic episodes, regionally uniform calcareous rocks of late Anisian age blanketed most of the outcrop area. These rocks contain a rich pelagic fauna and were evidently deposited below wave base. The more western of these rocks grade upward into dark cherty limestone which was deposited in basinal and slope environments concomitantly with uplift and erosion, during Ladinian time, in the central part of the Star Peak outcrop area. South and southeast of the uplifted area, subsidence and peritidal deposition took place followed by deposition of supratidal algal-laminate dolomite. This dolomite was deposited across the beveled uplift and correlates with platform-margin and basinal limestone farther west. Thereafter, lower Karnian platform limestone thickly blanketed the outcrop area and built regressively westward out over the basinal deposits. Following a widespread mid-Karnian break in deposition, local erosion, and deposition of terrigenous clastic rocks, platform carbonate rocks again blanketed the outcrop area during late Karnian time and extended an unknown distance westward, perhaps as far as the original site of the Sierran–eastern Klamath belt.

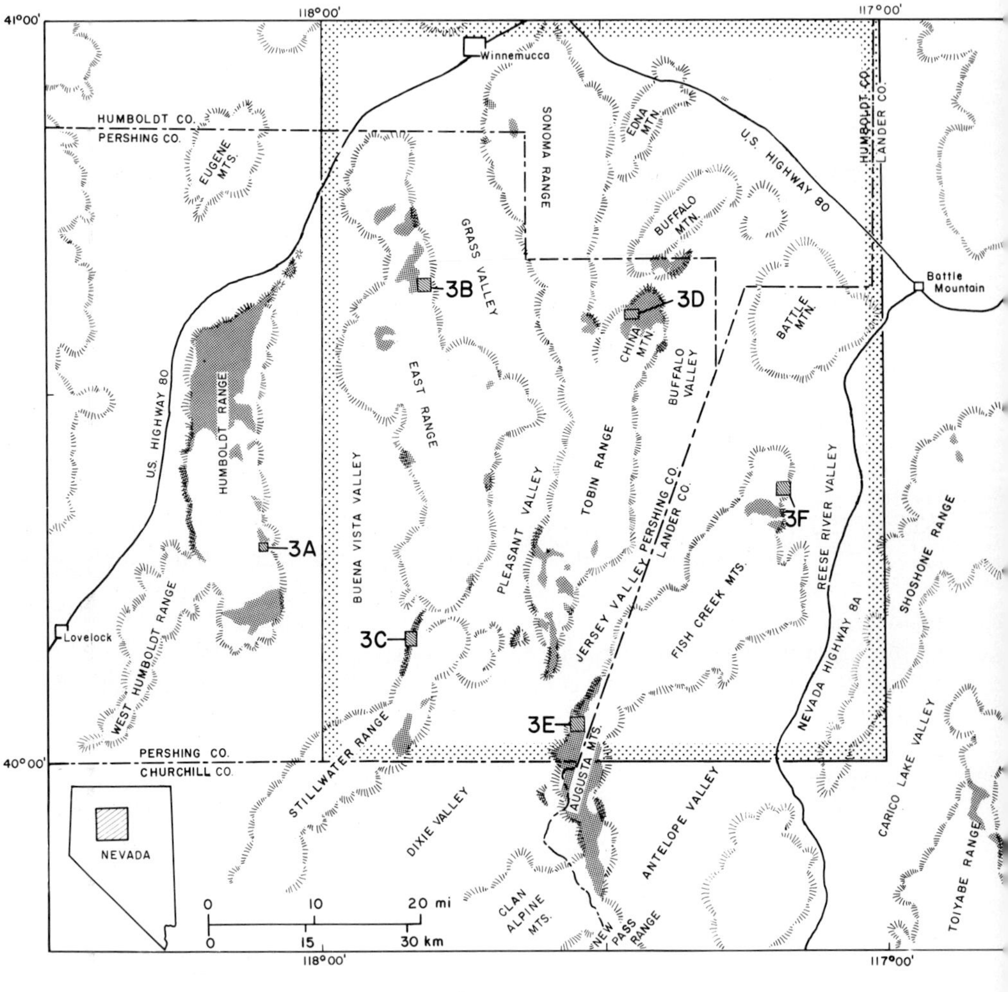

Figure 1. Index map showing location of Star Peak Group exposures (heavy stipple) in and near the Sonoma Range one-degree quadrangle (outlined by lightly stippled border). Location of maps shown in Figure 3 are indicated by labeled rectangles.

Introduction

The Star Peak Group is a distinct natural grouping of largely marine calcareous strata that range from Early to middle Late Triassic age and that occur in scattered exposures within an area exceeding 5,000 km^2 in northwestern Nevada (Fig. 1). Where the group is completely preserved, it attains thicknesses exceeding 1,000 m. The genetically coherent carbonate platform complex represented by the Star Peak Group contrasts sharply with the overlying Auld Lang Syne Group which consists of metapelitic rocks and sandstone. The Star Peak also contrasts with the unconformably underlying Koipato Group, which is composed of volcanic and clastic sedimentary rocks.

The Star Peak Group was first recognized as a major natural subdivision among the pre-Tertiary strata of northern Nevada more than 100 years ago during the initial exploration of this region by the 40th Parallel Survey under the direction of Clarence King. Since then, much knowledge about the depositional and tectonic history of the Star Peak Group has accumulated. The main obstacles to a coherent regional depositional synthesis of these rocks have been their discontinuous outcrops, difficulty in ascertaining the extent to which they have been rearranged by thrust faulting, and misinterpretation of massive secondary dolomite units in the lower part of the section.

All of the major components of the Star Peak Group are explicable in terms of a complicated pattern of carbonate deposition and diagenesis and contemporaneous tectonism. The regional synthesis presented in this paper describes an unusually compact example of lateral and vertical interrelations representing depositional environments that range from supratidal to below wave base (Fig. 2). Geologic maps of the Star Peak Group exposures that are significant for stratigraphic nomenclature and correlation are shown in Figure 3; a fence diagram illustrating the stratigraphic relations among these and other exposures of the group is shown in Figure 4. The Triassic carbonate units consist of diverse rock types that are best understood in terms of the environments in which they were deposited. Accordingly, a new scheme of stratigraphic nomenclature is proposed herein based on revision and documentation of the ages, correlations, and stratigraphic relations of the Star Peak components (Fig. 5).

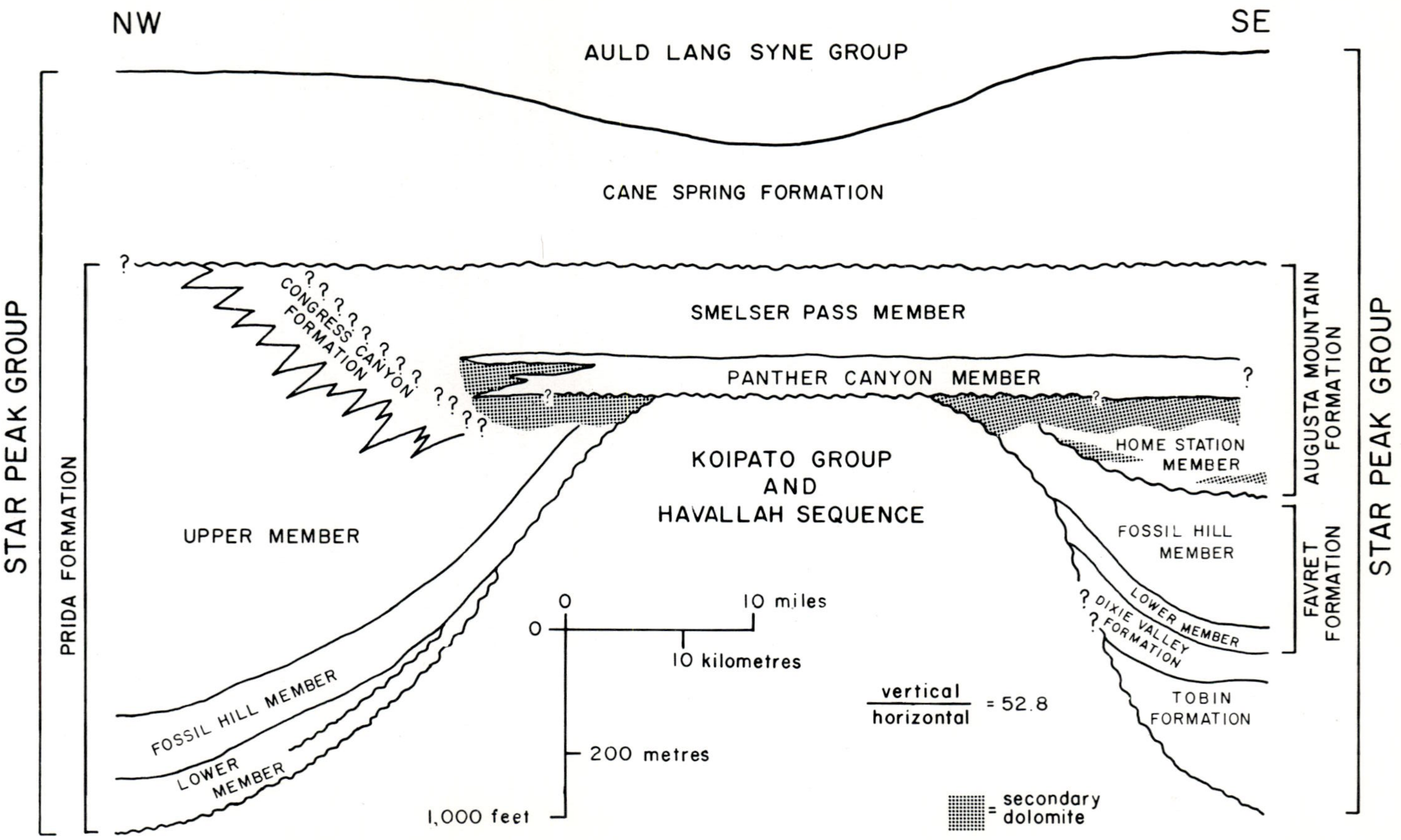

Figure 2. Diagrammatic cross section of the Star Peak Group trending southeasterly from the northern Humboldt Range through the southern Tobin Range to the Augusta Mountains.

Although the correlations, stratigraphic relations, and rank of previously named formations of the Star Peak Group have been revised, their lithologic boundaries have not been changed from those embodied in their original descriptions. Our revision has entailed incorporating some previously named units into others and dropping one name, the "Natchez Pass Formation."

REGIONAL GEOLOGIC SETTING

As first described by King (1878), typical exposures of the Star Peak Group are found in the vicinity of Star Peak in the Humboldt Range, where they were subsequently shown to be repeated structurally and to include fault blocks of the stratigraphically underlying Koipato Group. Nonetheless, the name "Star Peak" was correctly applied, in part, by King (1878) to isolated exposures of Triassic carbonate rocks elsewhere in the Humboldt Range and nearby ranges to the east. Subsequent study of the northern Humboldt Range by Cameron (1939) contributed to understanding the typical Star Peak section. This and work on the comparable Triassic rocks in the ranges to the east by Ferguson and others (1951a) enabled Silberling and Wallace (1967, 1969) to piece together the stratigraphic section of the type Star Peak.

For strata that are now regarded as parts of the Star Peak Group in the Sonoma Range one-degree quadrangle (Fig. 1), Muller and others (1951; see also Ferguson and others, 1951a, 1951b, 1952) established many new formations in order to express both lateral variation and the differences between sections that were interpreted as having been juxtaposed by tens of kilometres of displacement on the inferred "Tobin thrust" (Muller, 1949). So as not to prejudge the magnitude of this inferred thrust, Muller's supposed "Upper Plate (eastern) facies" and "Lower Plate (western) facies" of Triassic rocks in the Sonoma Range quadrangle were informally designated, respectively, as the "Augusta sequence" and "Winnemucca sequence" by Silberling and Roberts (1962).

Detailed work by Burke (1970b, 1973) in the southern Tobin Range, through which the typical trace of the "Tobin thrust" was supposed to extend, demonstrates that no such thrust fault exists and that the differences between the "Winnemucca" and "Augusta" sequences can be explained for the most part by rapid facies changes. Moreover, other recent findings regarding the character and age of thrust faulting in north-central Nevada (Silberling, 1970, 1973; MacMillan, 1971, 1972; Nichols, 1971; and Speed, 1971a, 1971b) indicate that the Star Peak Group rocks were not drastically rearranged by regional thrusting. The Golconda thrust, which does juxtapose entirely different upper Paleozoic facies in this region and was regarded originally as the same structure as the "Tobin thrust," evidently predates

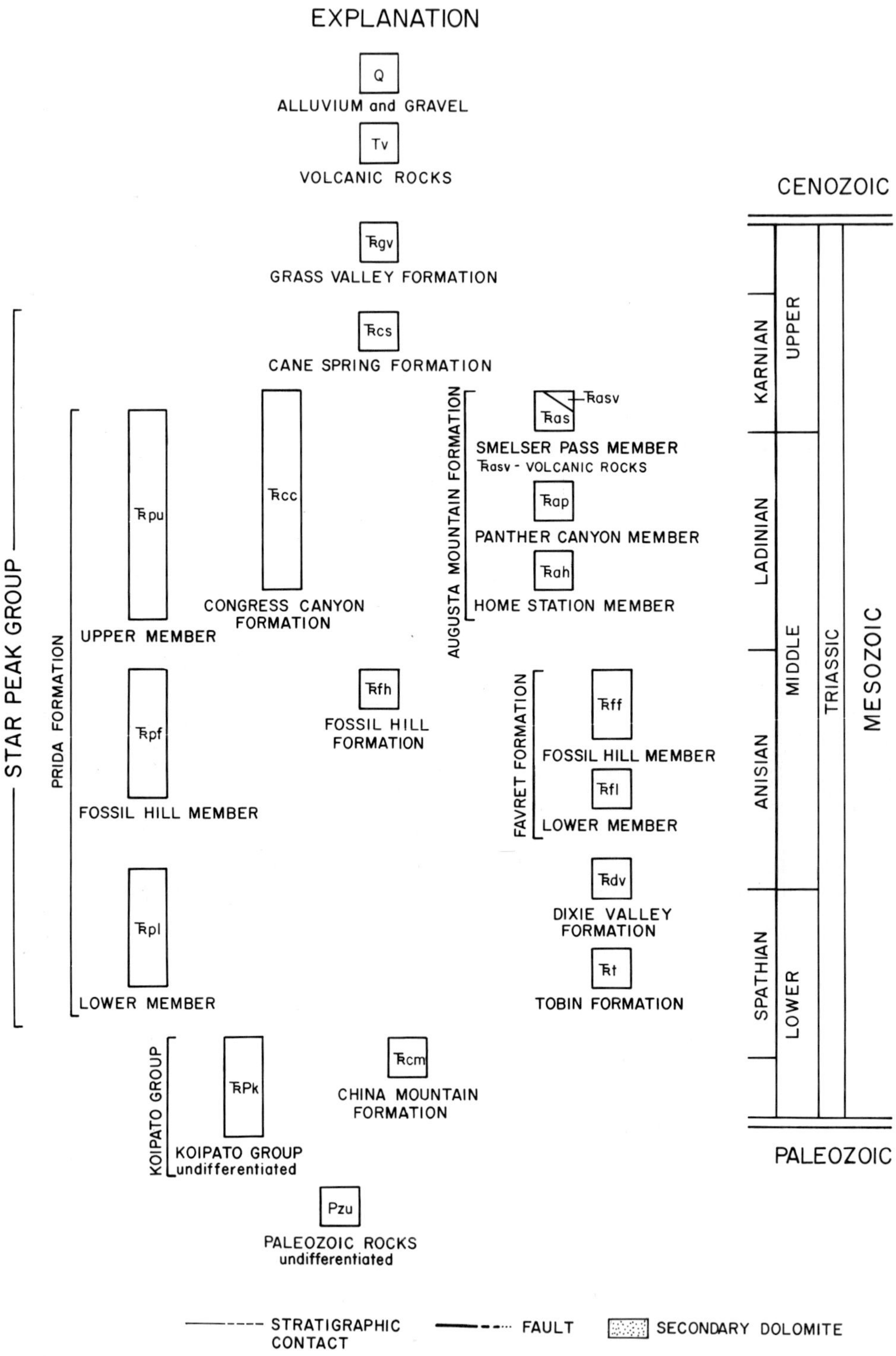

Figure 3 (facing pages). Geologic maps of selected exposures of the Star Peak Group in A, southern Humboldt Range; B, northern East Range; C, northern Stillwater Range; D, western

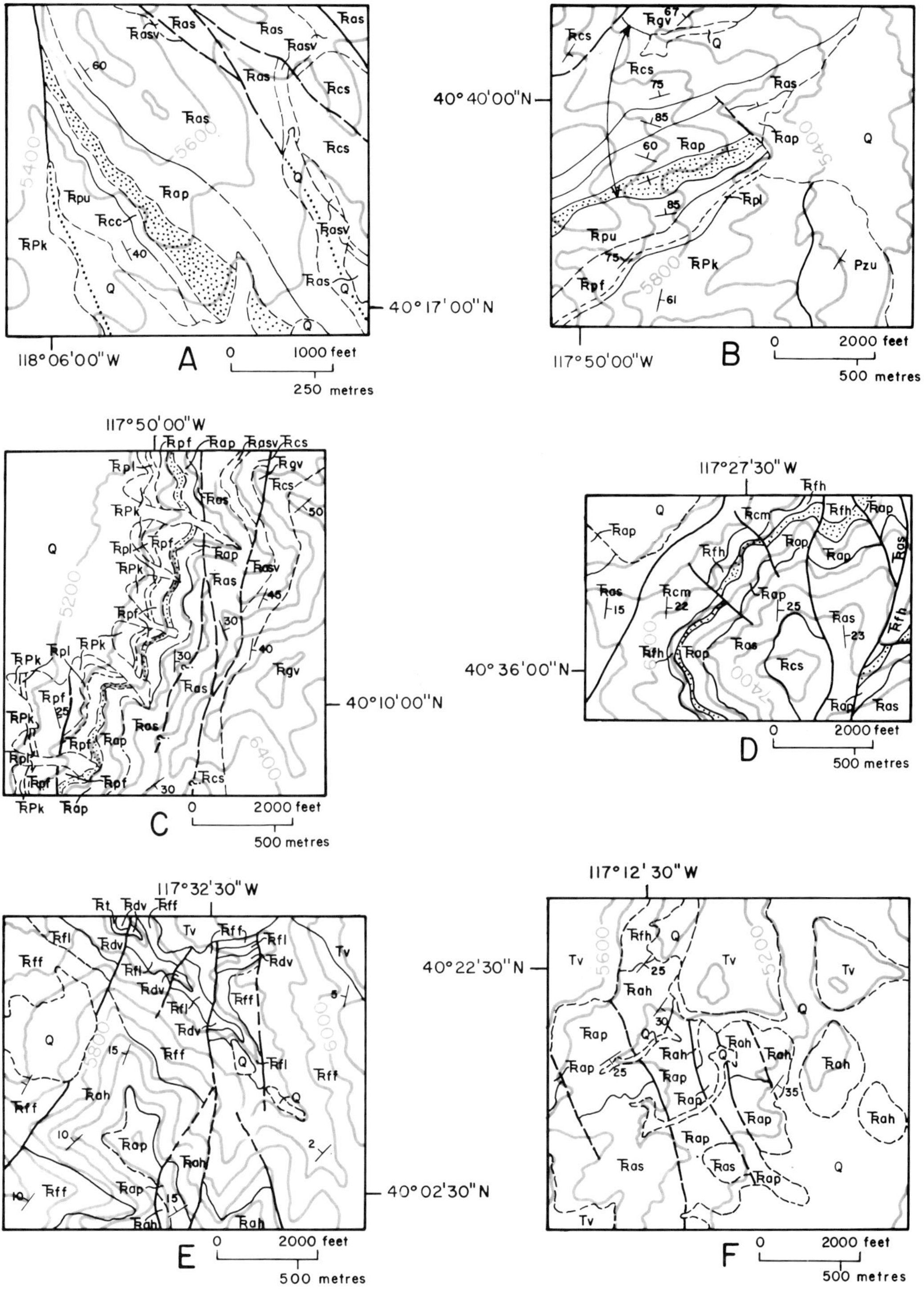

China Mountain; E, northern Augusta Mountains; F, northern Fish Creek Mountains. See Figure 1 for locations. Arrow on map B delineates rocks included in the "Natchez Pass Formation" of Ferguson and others (1951a).

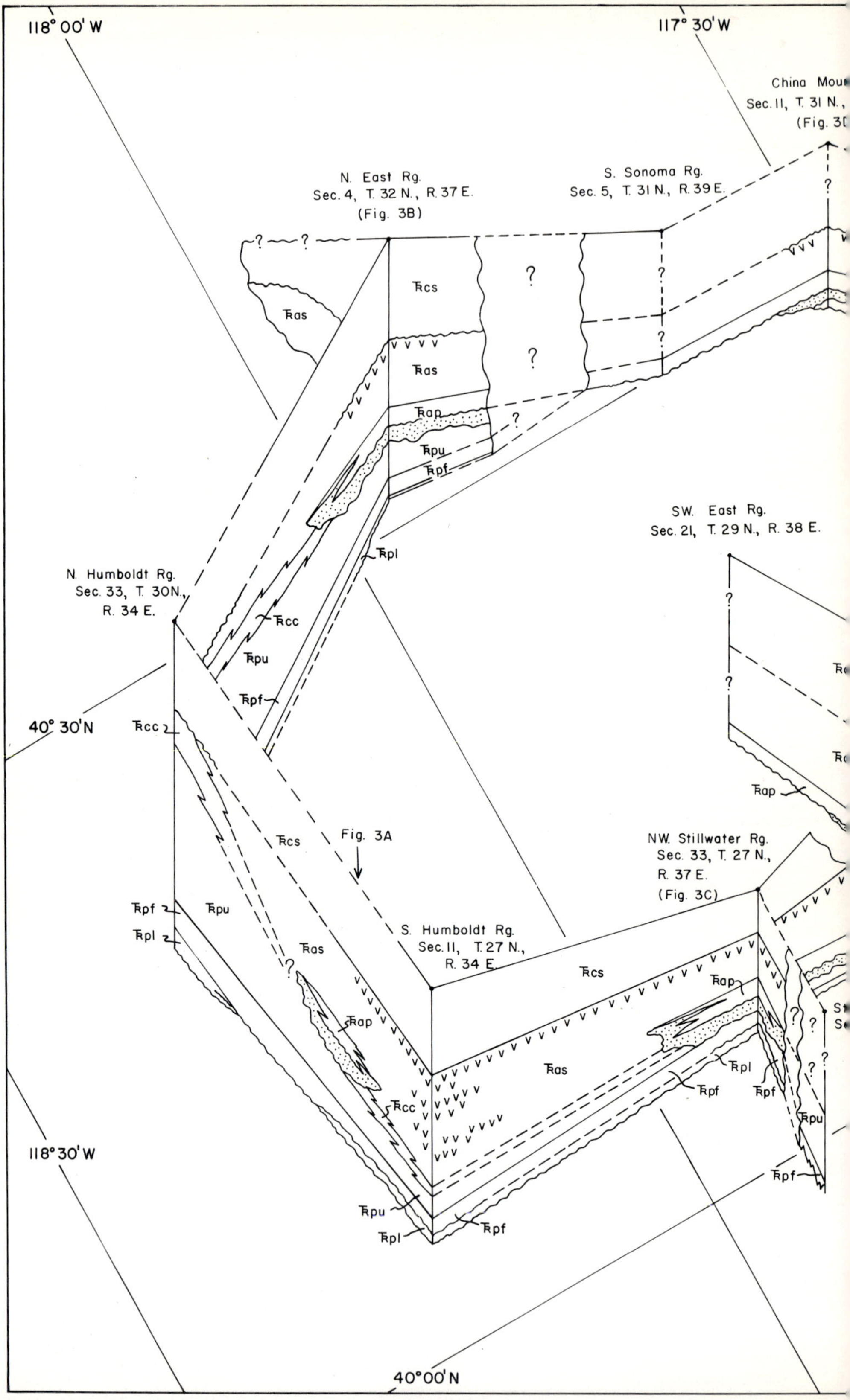

118° 00' W
117° 30' W
China Mou
Sec. 11, T. 31 N.,
(Fig. 3
N. East Rg.
Sec. 4, T. 32 N., R. 37 E.
(Fig. 3B)
S. Sonoma Rg.
Sec. 5, T. 31 N., R. 39 E.
Ŧcs
Ŧas
Ŧap
Ŧpu
Ŧpf
Ŧpl
Ŧcc
SW. East Rg.
Sec. 21, T. 29 N., R. 38 E.
N. Humboldt Rg.
Sec. 33, T. 30N.,
R. 34 E.
40° 30' N
Fig. 3A
NW. Stillwater Rg.
Sec. 33, T. 27 N.,
R. 37 E.
(Fig. 3C)
S. Humboldt Rg.
Sec. 11, T. 27 N.,
R. 34 E.
118° 30' W
40° 00' N

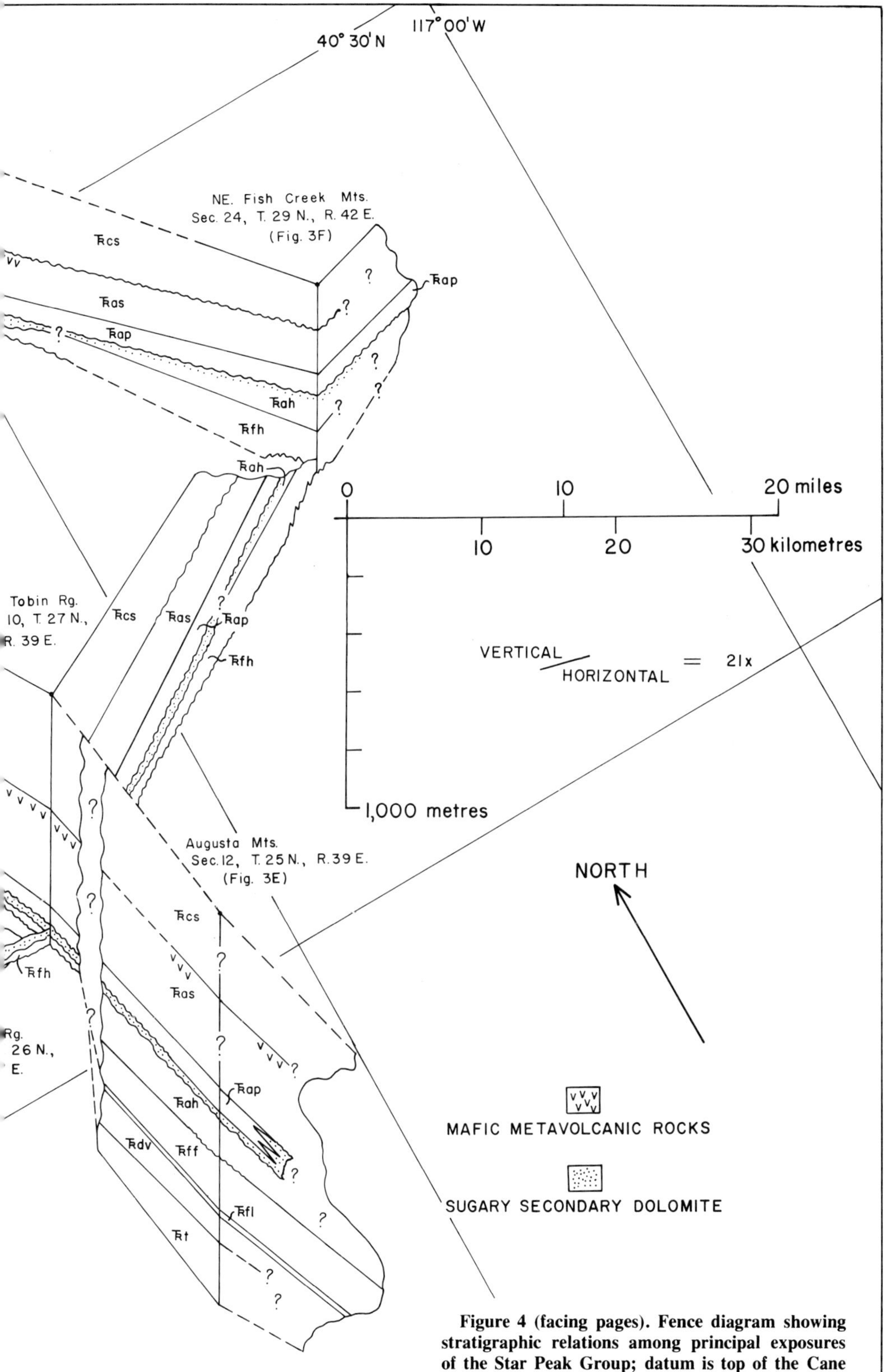

Figure 4 (facing pages). Fence diagram showing stratigraphic relations among principal exposures of the Star Peak Group; datum is top of the Cane Spring Formation. Formation symbols are the same as those in Figure 3.

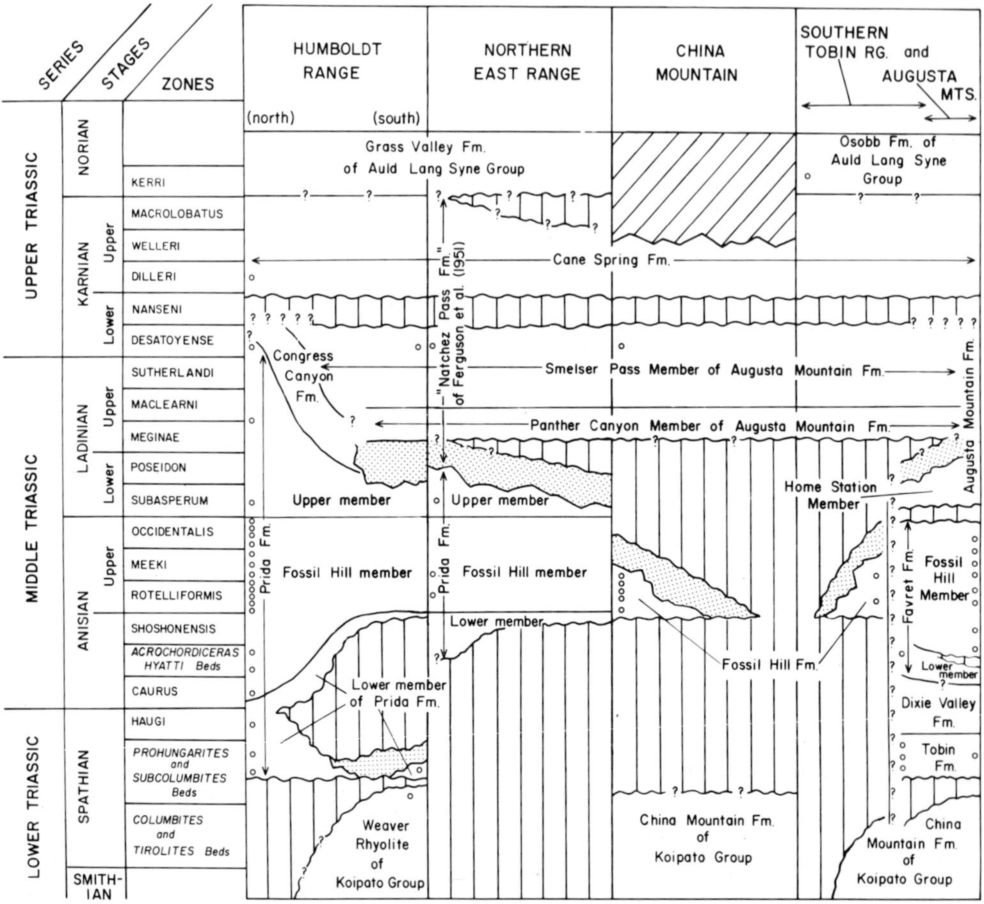

Figure 5. Time-stratigraphic correlation chart of the Star Peak Group at localities significant for stratigraphic nomenclature. Small circles represent occurrences of age-diagnostic fossils; stippled pattern indicates secondary dolomite; vertical ruling indicates stratigraphic hiatus, and diagonal ruling indicates lack of data.

deposition of the Star Peak Group, part of which overlaps the Golconda thrust (Nichols, 1971; Silberling, 1973, 1975). Displacement on the Golconda thrust in north-central Nevada most likely took place during the Sonoma orogeny in latest Permian or Early Triassic time. Therefore, lateral changes in Triassic rocks deposited subsequent to the Sonoma orogeny must be explicable without calling upon large-scale tectonic rearrangement, and separation of the Triassic strata into the tectonically delimited ''Winnemucca'' and ''Augusta'' sequences is no longer useful. Instead, these rocks lend themselves to treatment as a single lithologic entity whose internal variations (see Fig. 4) reflect a coherent, albeit complex, pattern of sedimentary environments. The lateral relationships do indicate, however, the possibility of some tectonic telescoping of facies on low-angle faults within the Star Peak Group, particularly in the three places in the fence diagram indicated by gaps across which lithologic correlations are questioned. These localities are in the northern Stillwater Range, southern Tobin Range, and between the East and Sonoma Ranges.

The Star Peak Group either rests unconformably on rocks of the predominantly volcanic Koipato Group (Fig. 2) which overlaps the Sonoma orogenic belt or, where the Koipato was removed by erosion, the Star Peak rests directly on the deformed Lower Permian and older(?) rocks of the Golconda allochthon. Deformation during the Sonoma orogeny is well shown at the south end of China Mountain (Fig. 6). Here the Hoffman Canyon thrust separates two parts of the intensely deformed Golconda allochthon having contrasting tectonic fabrics, and the thrust is overlapped by the nearly flat-lying tuffaceous rocks of the Koipato Group (Nichols, 1972). The Star Peak carbonate rocks in turn overlie the Koipato.

All of the known exposures of the Star Peak Group and closely related

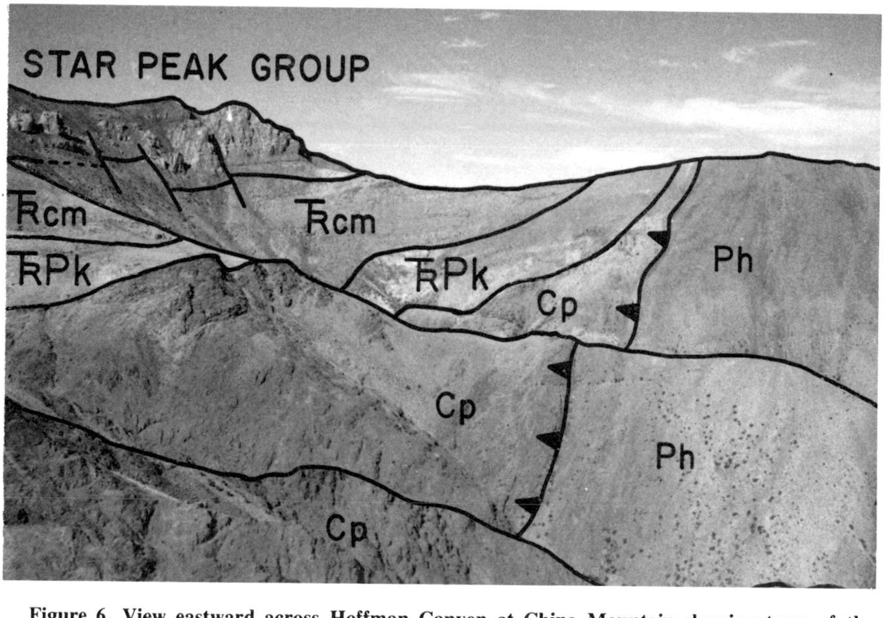

Figure 6. View eastward across Hoffman Canyon at China Mountain showing trace of the Hoffman Canyon Thrust, the Koipato Group, and part of the Star Peak Group. Rock-unit symbols correspond to those in Figure 3, except for the Golconda allochthon wherein Ph denotes the "Havallah Formation" and Cp the "Pumpernickel Formation," both after Ferguson and others (1952). Photo by S. W. Muller, about 1935.

strata are confined to the Humboldt Range and the ranges within the Sonoma Range quadrangle, except for isolated outcrops in the New Pass, Toiyabe, and Shoshone Ranges. Rocks belonging to the group have so far received only limited description in the form of map explanations, student theses, and a few geographically restricted studies. These rocks had not previously been considered as a genetically coherent group having wide regional extent. Some attempt at interpreting the regional depositional framework of the Star Peak Group in the Humboldt Range was made by Silberling and Wallace (1969) who recognized in the type Star Peak exposures the same formational units that were defined by Ferguson and others (1951a) in the East Range, the next range to the east. Geographically restricted research by Burke (1973) in the southern Tobin Range, by Nichols (1972, 1974) at China Mountain, and by the Stanford Geological Survey summer field camps in the Augusta Mountains during 1970 and in the northern East Range during 1971 improved knowledge of the distribution and correlation of previously recognized subdivisions within the group in the vicinity of the Sonoma Range quadrangle. Farther east, near Hall Creek in the northern Toiyabe Range, Stewart and McKee (1977) discovered a large exposure of Triassic carbonate and clastic rocks that resemble part of the Star Peak Group. Rocks exposed in small areas in the northern Shoshone Range were compared by Gilluly and Gates (1965) with the China Mountain Formation but are now recognized as part of the Star Peak (Nichols, 1971). In the New Pass Range, south of the Sonoma Range quadrangle, Triassic rocks resembling those of the Star Peak Group have been studied by MacMillan (1972) and Willden and Speed (1974).

A major obstacle to an understanding of the Star Peak Group has been the massive bodies of saccharoidal dolomite, locally forming much of the lower part of the section, which have been misinterpreted as primary stratigraphic units. The secondary and nonstratigraphic character of these dolomite bodies has been demonstrated by Nichols (1974) who found that they relate spatially to an overlying unit, the Panther Canyon Member of the Augusta Mountain Formation (Fig. 2). The Panther Canyon is a fine-grained, laminated, stromatolitic dolomite which evidently formed in a sabkha-flat environment. This distinctive unit appears in all exposures of the Star Peak Group, except those in parts of the Humboldt and Stillwater Ranges, and is coextensive with the underlying secondarily dolomitized rocks.

Hydrothermal alteration of the carbonate rocks of the Star Peak Group, especially in the more northwestern outcrops, seriously hampers petrographic interpretation of depositional and diagenetic fabrics. Such alteration is not surprising, because these exposures occur along the eastern edge of the batholithic region in northwestern Nevada, interpreted by Smith and others (1971) as connecting the Sierra Nevada and Idaho Batholiths.

TERMINOLOGY AND CONVENTIONS

Most of the carbonate rocks of the Star Peak Group have undergone some degree of hydrothermal recrystallization, but insofar as possible the terminology used in their petrographic description is based on a combination of Folk's (1959, 1962) and Dunham's (1962) classification of carbonate rocks. Colors are noted by comparison with the Rock Color Chart (Goddard and others, 1948). Terminology for thickness of bedding and parting follows Ingram (1954) and the term "massive" is applied to rocks characterized by relatively thick beds that lack apparent internal compositional variations or structures.

Stratigraphic thicknesses are calculated in most cases from geologic maps at scales of 1:24,000 or larger. In view of the tectonic complexity of the region, we believe that these are more accurate approximations than would be attained from a few measured sections.

The time scale used for strata of the Star Peak Group is that proposed by Silberling and Tozer (1968), as modified by Tozer (1971). The usage of stage names and placement of their boundaries conform to generally accepted conventions, except that the Anisian-Ladinian interstage boundary is now drawn at a somewhat lower level by some European authors (for example, Kozur, 1974). The names of ammonite zones are written in the form used by Silberling and Tozer (1968).

Stratigraphy

The Star Peak Group is as much as 1,200 m thick and ranges in age from late Early to middle Late Triassic. The rocks display a complicated pattern of carbonate and clastic-rock sedimentation which resulted in part from contemporaneous tectonism. The succession, lateral relations, and thickness variations of rock units are illustrated diagrammatically in Figure 2 and shown by a fence diagram in Figure 4. The age relations of these rocks are given in Figure 5, and the evolution of the carbonate platform that much of the group represents in portrayed in Figure 31. Aside from being composed mainly of carbonate rocks, the Star Peak Group differs from the concordantly underlying Kiopato Group in that the Star Peak lacks evidence of silicic volcanism. In contrast, even the least volcanic parts of the Koipato, such as the predominantly clastic China Mountain Formation, contain first-cycle rhyolitic tuff. Throughout its outcrop area, the Star Peak Group is overlain by the Auld Lang Syne Group, of late Late Triassic age, which consists of sandy and argillaceous strata intercalated with some calcareous rocks. The initiation of Auld Lang Syne sedimentation is marked by the abrupt appearance of fine-grained clastic rocks that include micaceous feldspathic sandstone, a rock type not found in the Star Peak Group.

In revising the stratigraphic nomenclature, we have exercised the philosophy that names for rock units should be defined strictly on the basis of lithologic character, thereby reflecting patterns of depositional environments that changed both laterally and vertically. Because of marked lateral stratigraphic changes in the Star Peak Group, especially in the lower portion, several different formational names are needed for correlative units in geographically different parts of the group. Therefore, the strata of the Star Peak Group are discussed first in the eastern region of Star Peak outcrops and then in the central and western regions.

EASTERN REGION

The strata that characterize the older parts of the Star Peak Group in the eastern region are, for the most part, restricted to the southernmost

Tobin Range and the Augusta Mountains. Some of these rocks also crop out in the northeastern Fish Creek Mountains, where the base of the section is not exposed. These strata, which belong to the Tobin, Dixie Valley, and Favret Formations and the Home Station Member of the Augusta Mountain Formation, were deposited to the south and east of an area that was affected contemporaneously by uplift and erosion during Middle Triassic time. They display a variability in thickness and lithologic character that does not characterize the younger, post-Ladinian parts of the Star Peak Group, which were deposited across the locally uplifted areas. The younger Triassic rocks in the eastern region are assigned to the Panther Canyon and Smelser Pass Members of the Augusta Mountain Formation and to the Cane Spring Formation (Fig. 2). The thickness and lithology of these two younger members of the Augusta Mountain Formation remain essentially constant throughout the eastern and central regions. In the Humboldt Range these two members interfinger with the Congress Canyon Formation. The Cane Spring Formation is a regionally widespread unit whose character does not vary significantly throughout the Star Peak region. Therefore, descriptions of the younger parts of the Star Peak in the eastern part generally apply to the entire Star Peak area.

The lower part of the Star Peak Group in the eastern region is represented by the Tobin Formation and the overlying Dixie Valley Formation which range in age from latest Early (late Spathian) to earliest Middle Triassic (early Anisian). These two units form only a minor part of the Star Peak Group, because they are geographically restricted. Aside from exposures in the Tobin Range, Augusta Mountains, and a small area in the southwestern East Range, possible correlatives of the Tobin and Dixie Valley Formations crop out in the Wild Horse district in the low hills connecting the Augusta Mountains and New Pass Range (Fig. 1). Farther south, in the New Pass Range, the thick section of coarse clastic strata that separates highly deformed rocks of the Golconda allochthon from calcareous beds of Anisian age is regarded by us as equivalent to the China Mountain Formation of the Koipato Group. Willden and Speed (1974), however, regard the upper, mainly conglomeratic part of this section as a possible lateral equivalent of the Tobin and Dixie Valley Formations.

At China Mountain clastic rocks that rest unconformably on either volcanic rocks of the Koipato Group or the Havallah sequence were designated originally as the China Mountain Formation by Ferguson and others (1952). They regarded these as a thinner and coarser clastic northern equivalent of the Tobin and Dixie Valley Formations. The unfossiliferous China Mountain Formation was thus regarded as being distinctly younger than the Koipato. However, the China Mountain Formation has now been recognized over a wider area, and is considered as part of the Koipato Group (Nichols, 1971) because it underlies the Tobin Formation of the

Star Peak Group in the southern Tobin Range (Burke, 1973) and includes siliceous tuffaceous rocks. Throughout its extent the China Mountain is probably older than any part of the Star Peak Group and may be correlative with interstratified clastic and volcanic rocks that form the upper part of the Koipato Group in the Humboldt Range and contain ammonite faunas of middle Spathian age (Silberling, 1973).

Tobin Formation

The Tobin Formation (Muller and others, 1951), which forms the base of the Star Peak Group in the eastern region, consists principally of terrigenous fine-grained clastic rocks which are more or less calcareous and include some limestone and coarse clastic rocks. In its typical exposures in the southernmost Tobin Range, where it has been studied in detail by Burke (1973), the Tobin Formation is about 150 m thick and is mostly dark, unfossiliferous, commonly laminated, subfissile or irregularly fractured, silty mudstone. Nodular or concretionary interbeds of either crinoidal or dense impure limestone occur sporadically throughout the section. Limestone interbeds are most abundant near the base where they contain upper Spathian *Subcolumbites* ammonite faunas such as those described by Kummel (1969) and listed by Burke (1973, p. 40). The calcareous and fine-grained clastic strata are separated from the underlying Koipato Group in most places by as much as a few metres of calcareous or dolomitic gritty sandstone. The basal conglomerate and mafic volcanic rocks, originally included by Muller and others (1951) in the lower part of the Tobin, are shown by Burke (1973) to underlie unconformably the main body of the formation and are better assigned to the Koipato Group. In its typical exposures the top of the Tobin Formation is a distinctive unit, 25 m thick, of medium- to thick-bedded massive and mottled silty limestone which contains a diverse molluscan fauna whose preservation suggests early diagenetic dissolution of aragonitic shells. Poorly preserved ammonites from this limestone unit include "*Prohungarites* sp. indet." of Kummel (1969, p. 552) which is probably representative of the higher parts of the "*Subcolumbites* beds" of Silberling and Tozer (1968). *Lingula* shells are conspicuous in this uppermost limestone unit as well as in the lower part of the formation.

In the Augusta Mountains the Tobin Formation exceeds 200 m in thickness (Fig 4). The clastic rocks here, which are finer grained than those in the Tobin Range, are mostly finely fractured dark argillite. Calcareous pyritic concretions in places contain ammonites, pectenacid bivalves, and ichthyosaurian bones, but few other kinds of fossils are preserved in these rocks. The basal few tens of metres of the Tobin here are calcareous sandstone. The silty limestone unit that occurs at the top of the type Tobin Formation is represented in the Augusta Mountains by conglomerate interstratified with massive coarse-grained crinoidal limestone.

The preponderantly dark, carbonaceous and pyritic character of the Tobin mudstone, its fine lamination, and absence of fossils of benthonic organisms other than *Lingula,* suggest quiet-water deposition of much of the formation in a restricted shallow marine basin under euxinic conditions. However, the occurrence of ammonites and crinoidal limestone, especially in the Tobin Range, indicates proximity with more open-marine environments. A change to well-aerated and more energetic conditions of sedimentation, and to diagenesis affected by meteoric waters, is indicated by the character of the silty limestone unit at the top of the formation.

The original extent of the Tobin Formation is difficult to ascertain. Its abrupt disappearance northward within a northwesterly trending zone in the southern Tobin Range is part of the peculiar map pattern that led Muller and others (1951) to hypothesize the "Tobin thrust" juxtaposing different facies of Triassic rocks. As observed by Burke (1973), the more northern exposures of the Tobin Formation in the southern Tobin Range become progressively more silty and contain more abundant stringers and beds of grit and pebble conglomerate derived from the Koipato Group. Extrapolating this trend, the Tobin may be represented by about 120 m of coarsely cross-bedded calcareous sandstone and conglomerate that directly underlies the Fossil Hill Formation, of late Anisian age, about 4 km northeast of the closest known outcrop of the Tobin Formation. This sandstone and conglomerate would therefore mark the original edge of the Tobin basin. However, it could equally well be part of the China Mountain Formation of the Koipato Group. The Dixie Valley Formation, part of the Favret Formation, and the Home Station Member of the Augusta Mountain Formation also disappear within the same northwesterly trending zone where the Tobin Formation disappears. Only pre-Star Peak and Cenozoic rocks are exposed in this zone. Therefore, juxtaposition of different Triassic sections by fault displacement remains a possibility, even though Burke's careful mapping of the stratigraphically and structurally complex rocks of the Koipato Group and the Golconda allochthon failed to identify an obvious trace for a fault between the contrasting Triassic sections. The existence of the "Tobin thrust," as hypothesized by Muller and others (1951), can be ruled out, however, because the Panther Canyon Member of the Augusta Mountain Formation and all younger parts of the Star Peak Group extend with little change across this hypothetical structure. Also, the strata directly overlying the Star Peak Group in the Tobin Range and originally assigned to the Grass Valley Formation, which is part of the "lower-plate facies" of the Tobin thrust according to Muller and others (1951), lithologically should be regarded as belonging to the Osobb Formation (Burke, 1973). Typical exposures of the Osobb in the Augusta Mountains would belong to the upper plate of the postulated "Tobin thrust."

The Tobin Formation may have extended far to the west and then was eroded away during uplift in early Anisian time. Reasons for entertaining this possibility are discussed further in the section on the western region. The lower member of the Prida Formation in the western region is in part synchronous with the lower part of the Tobin and, where not greatly altered by dolomitization, is lithologically similar to it.

Dixie Valley Formation

The Dixie Valley Formation (Muller and others, 1951) is coextensive with the Tobin Formation in the eastern region cropping out along the northwestern flank of the Augusta Mountains—the type locality of the formation—and in the southern Tobin Range. It is a distinctive unit composed of interbedded yellow and brown conglomerate, sandstone, and dolomite and maroon siltstone beds which form colorful outcrop bands. The formation, which ranges from 100 to 200 m in thickness, does not lend itself to a generalized stratigraphic description, as the clastic rocks and dolomite that make up the section vary widely in proportions and character.

In the Tobin Range the basal beds that rest on a well-defined and distinctly channeled surface on the Tobin Formation, are conglomeratic, showing great variability in grain size and shape and bedding characteristics (Burke, 1973). The rest of the section comprises interbedded dolomite, dolomitic siltstone and sandstone, and lesser amounts of poorly sorted pebble conglomerate. The dolomite is more or less silty, commonly laminated, and yellow weathering, the siltsone and sandstone exhibit mud cracks and current ripple marks, and the conglomerate occurs in lenses and channel fills which have high-angle cross bedding and pebble imbrication. Current directions, based on cross bedding in conglomeratic units and ripple marks on siltstone beds, are generally to the south and southwest (Burke, 1973). In the Augusta Mountains, pebbly dolomitic sandstone and silty dolomite beds at the base of the Dixie Valley Formation are not so sharply distinguished from the underlying Tobin Formation. Higher parts of the section here resemble that in the Tobin Range, and imbrication of pebbles in the Augusta Mountains indicates deposition of the coarser clastic rocks by southeasterly flowing currents.

The clastic rocks of the Dixie Valley Formation indicate alluvial, nonmarine deposition. Moreover, sedimentation near sea level is indicated because they are interstratified and mixed with algal-laminate, desiccated dolomite suggestive of a marine-influenced supratidal depositional environment. In general, the formation represents shoreline deposition associated with contemporaneous uplift and erosion within part of the Star Peak outcrop area to the north. Fossils are lacking in the Dixie Valley, but its age is within the interval between latest Early Triassic (late Spathian) and early Middle Triassic (middle Anisian) time, based on fossils in underlying and overlying strata.

Favret Formation

Rocks assigned to the Favret Formation, like those of the Tobin and Dixie Valley Formations, are limited to the Augusta Mountains and the southern Tobin Range. The name "Favret" was applied originally by Muller and others (1951) to calcareous Middle Triassic strata that crop out on the south wall of Favret Canyon in the Augusta Mountains. We divide this formation into two members: an informal, unnamed lower member and an upper member, herein named the "Fossil Hill Member" whose typical exposures are to the northeast in the Humboldt Range. Part of Favret Canyon is depicted in Figure 3E showing the location of representative exposures of these two members and their relations to stratigraphically lower and higher parts of the Star Peak Group. The name "Fossil Hill" is herein applied (Fig. 2) to rocks throughout the Star Peak outcrop area, because strata assigned to this member form a laterally uniform unit of offshore calcareous rocks containing a distinctive, exclusively pelagic Anisian cephalopod and bivalve fauna. This unit, composed of interstratified calcareous mudstone and black limestone, undergoes minor lateral changes in composition, but its character and the age of its youngest beds remain relatively constant throughout a region exceeding 5,000 km^2. Significant lateral changes do occur, however, in the lower age limit of the Fossil Hill and in the age and character of the rocks beneath it. In the Augusta Mountains and part of the southern Tobin Range, the Fossil Hill Member rests on the lower member of the Favret.

The two members of the Favret Formation are easily distinguished from one another. The lower member is characterized by thick-bedded to massive, light-gray (5Y 8/1) limestone, which forms benchlike outcrops and attains a maximum thickness of about 30 m (Fig. 7). Pronounced low-angle, wedge-shaped cross stratification and large fragments of molluscan shells are conspicuous. As viewed in thin section, the limestone of the lower member is composed primarily of crinoid ossicles, rounded and micritized shell fragments, and some oolites set in a sparry calcite matrix (Fig. 8). These features, along with the well-sorted skeletal-grain fraction and the absence of silt- and clay-size components, are indicative of high-energy, shallow-water deposition. Data to support the more detailed interpretation of an offshore shoal edge or nearshore bar are lacking, as the lower member has a limited outcrop area.

At Favret Canyon the Fossil Hill member is a 200-m sequence of fissile to flaggy, dark-gray (N3) calcareous shale containing concretionary interbeds of fossiliferous lime mudstone that are most abundant at the top of the

Figure 7. Benchlike outcrops of the lower member of the Favret Formation at its type locality in Favret Canyon, Augusta Mountains. In this view the member is about 20 m thick.

unit. The member contains a rich fauna of cephalopods and pelagic bivalves. When broken, the chambers of ammonoids from concretions commonly yield liquid hydrocarbons. The more calcareous interbeds in the Fossil Hill Member (Figs. 9A, 9C, 9D) are composed of shells and fragments of thin-shelled bivalves, cephalopods, and ostracods which occur in a lime matrix with a small amount of finely disseminated quartz silt. The micritic texture and the absence of high-energy allochems suggest deposition below wave base, in contrast to the high-energy conditions interpreted for the lower member of the Favret Formation. A disconformity could exist between the two members of the Favret, but the amount of stratigraphic hiatus, if any, is unknown, because the lower member is devoid of biostratigraphically useful larger invertebrate fossils. Data are also lacking for determining the time gap at the erosion surface between the Favret Formation and the overlying Home Station Member of the Augusta Mountain Formation. Although the top of the Favret in the Augusta Mountains is an eroded surface, the upper part of the Fossil Hill Member here has not been altered by dolomitization as it has in the central Star Peak region.

In the Augusta Mountains the Fossil Hill Member contains ammonite and halobiid bivalve faunas of middle and late Anisian age. The basal beds of the member at Favret Canyon contain coarsely rugate *Posidonia*-like bivalves which elsewhere characterize the *Acrochordiceras hyatti* beds. Concretionary limestone beds containing the halobiid *Enteropleura bittneri* and ammonites of the Shoshonensis Zone occur about 40 to 50 m higher in the section. Upper Anisian *Daonella*-bearing faunas are represented almost continuously throughout the upper 100 m of the member, and the stratigraphically highest faunas belong to the uppermost Anisian Occidentalis Zone.

The lower part of the Fossil Hill Member and the lower member of

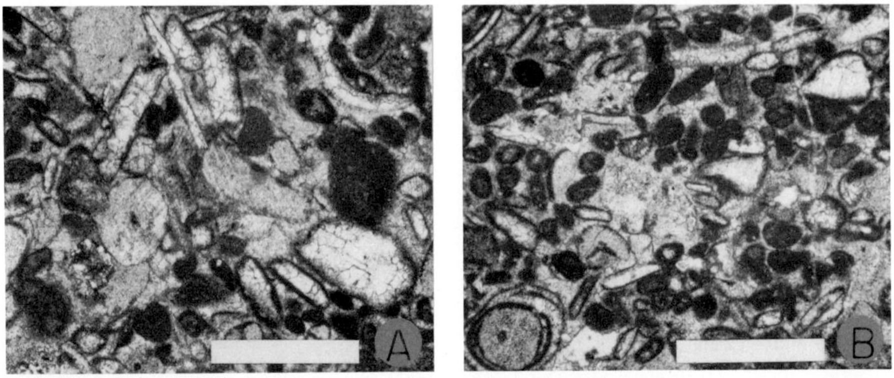

Figure 8. Photomicrographs of the lower member of the Favret Formation from Favret Canyon, Augusta Mountains. A, B, biosparite with micritized crinoid ossicles and rounded shell fragments. Bar = 0.50 mm.

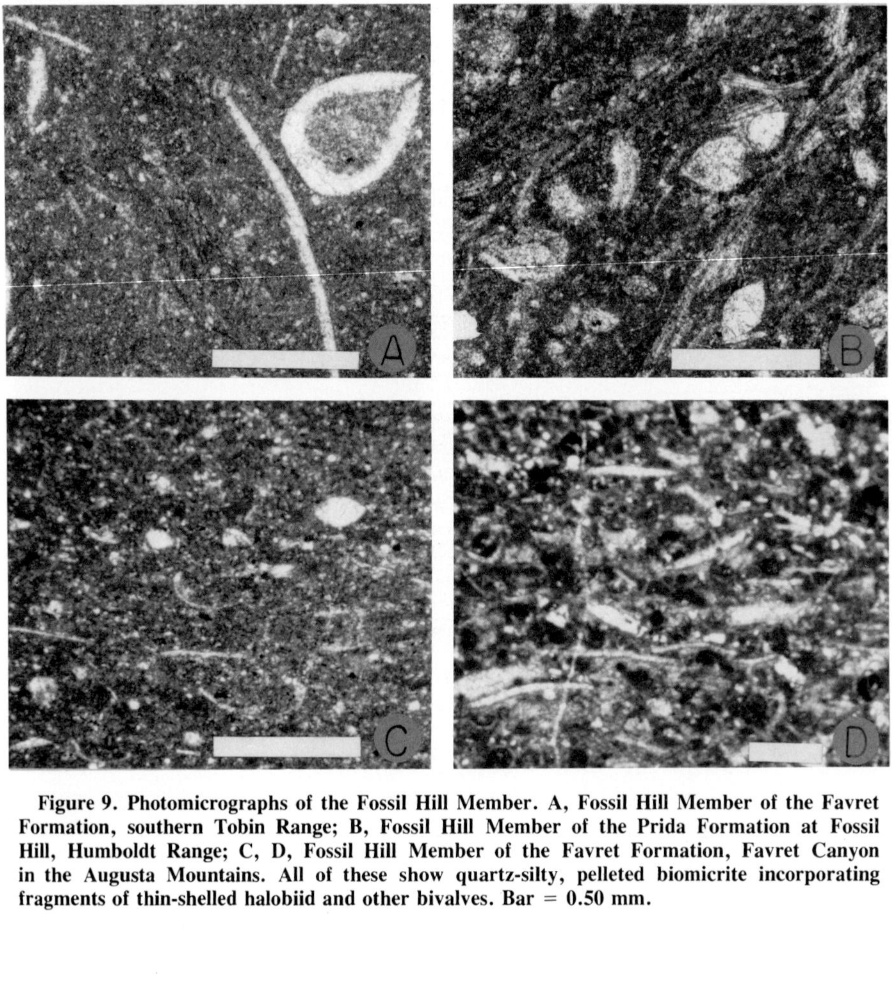

Figure 9. Photomicrographs of the Fossil Hill Member. A, Fossil Hill Member of the Favret Formation, southern Tobin Range; B, Fossil Hill Member of the Prida Formation at Fossil Hill, Humboldt Range; C, D, Fossil Hill Member of the Favret Formation, Favret Canyon in the Augusta Mountains. All of these show quartz-silty, pelleted biomicrite incorporating fragments of thin-shelled halobiid and other bivalves. Bar = 0.50 mm.

the Favret Formation disappear abruptly to the north within the southern Tobin Range. As discussed previously, all of the Tobin Formation, Dixie Valley Formation, and Home Station Member of the Augusta Mountain Formation also disappear within the southern Tobin Range, an aggregate loss of about 600 m of strata. Although telescoping of pre-Tertiary rocks on the hypothetical "Tobin thrust" demonstrably did not occur, some significant amount of fault juxtaposition of Triassic facies here does remain a possibility.

Augusta Mountain Formation

When Muller and others (1951) first applied the name "Augusta Mountain Formation" to the 750 m of Middle and Upper Triassic carbonate rocks exposed in the Augusta Mountains, they divided the formation into three informally named members. The Panther Canyon Formation was established as a separate formation and mapped as such at its type locality in the China Mountain area by Ferguson and others (1952). They also considered the gray limestone sequence overlying the Panther Canyon to be an attenuated northern equivalent of the entire Augusta Mountain Formation. At that time they did not realize that the middle member of the Augusta Mountain Formation, as defined by them, is, in fact, the Panther Canyon Formation (Nichols, 1972). This stratigraphic conflict has been resolved simply by lowering the rank of the Panther Canyon to that of a member within the Augusta Mountain Formation (Nichols, 1974). The originally unnamed middle member of the Augusta Mountain Formation thereby becomes the Panther Canyon Member; the lower and upper members are here named the Home Station Member and Smelser Pass Member, respectively.

Home Station Member. The Home Station Member consists of lithologically diverse and laterally variable dolomitic platform carbonate rocks that crop out in the Augusta Mountains, Fish Creek Mountains, and part of the southern Tobin Range. The name of the member derives from Home Station Ranch which is about 30 km northwest of the type section which is located on the ridge south of Favret Canyon. The geologic relations of the Home Station Member here are shown on Figure 3E. Ferguson and others (1951b) did not recognize separate members within the Augusta Mountain Formation in the Fish Creek Mountains, but the Home Station and younger members of the formation are mappable (Fig. 3F). The presence of the Home Station in the Fish Creek Mountains and its absence to the west at China Mountain reflects the development during Ladinian time of the structural high in the central Star Peak region (Figs. 2, 5).

Inner-platform strata, represented by interstratified calcareous or dolomitic limestone, algal-laminate fine-grained dolomite, and dolomitic or siliceous siltstone and sandstone, compose part of the Home Station Member. The unit also includes massive, cliff-forming saccharoidal secondary dolomite. These secondary dolomite bodies are interstratified with the primary sedimentary rocks and resulted from dolomitization of some of them. The secondary dolomites weather a monotonous light grayish white (5YR 7/1) and commonly contain chert blebs, stringers or lenses of sandstone, and wisps of chert grit or pebbles. Clastic material is coarser grained and more abundant in the Fish Creek Mountains than in the more southern exposures. Wisps of pebbles are also found in the light-gray (N6) to grayish-orange (10YR 7/4) weathering, medium-bedded, fine-grained and algal-laminate dolomicrite (Fig. 10). Bird's-eye and shrinkage features identify these dolomicrites as supratidal deposits. Minor amounts of brown-weathering (10R 5/4) calcareous, dolomitic or siliceous siltstone and sandstone are found throughout the section but seem to be concentrated in the upper portion of the unit. Where not completely altered by secondary dolomitization, a small amount of bedded gray limestone occurs and locally contains coarse skeletal debris.

At its type locality in the Augusta Mountains, the Home Station Member is about 250 m thick. The base is marked by dark-brown dolomitic sandstone

Figure 10. Light-colored algal-laminate dense dolomite having stringers and medium-thick interbeds of dark siliceous sandstone and pebble conglomerate. These rocks in the Home Station Member, Augusta Mountain Formation, are interlayered with coarsely crystalline secondary dolomite. northern part of the Fish Creek Mountains.

fillings 1 to 2 m thick in scour channels cut into the underlying Fossil Hill member (Fig. 11). These sandstone channel fillings are sharply separated by an undulant surface from spectacular intraformational breccia about 1 m thick. This in turn is overlain by yellowish-gray, fine-grained dolomite that contains wisps of sand and grit and stringers of chert. The disconformity at the base of the member is not recognizable in the Fish Creek Mountains, where the Home Station is about 75 m thick. The contact with the overlying Panther Canyon Member is marked by a subtle upward transition into dense massive dolomite which lacks coarse clastic and skeletal debris and interstratified saccharoidal dolomite.

In the type area the uppermost few tens of metres of strata beneath the Panther Canyon Member consist of massive sugary dolomite, at the base of which nodules or thin bedding-plane layers of dark secondary chert are concentrated. This "chert screen" suggests that here, as elsewhere, secondary alteration of the carbonate rocks immediately beneath the Panther Canyon Member resulted from solutions related to the Panther Canyon or the surface on which it was deposited. Nodules of secondary chert in the sugary dolomite (Fig. 11) just below the base of the Panther Canyon are composed of fibrous length-slow chalcedony which was regarded by Folk and Pittman (1971) as an index of replacement associated with solutions of evaporitic origin.

The interstratification of pebbly or algal-laminate dolomite with a variety of originally calcareous or fine-grained clastic rocks indicates deposition of the Home Station Member under fluctuating inner-platform marine and supratidal conditions. Age-diagnostic fossils have not been found in these rocks, and the age of this unit may vary laterally because of the unconformity at its base. The youngest fossils in the underlying Fossil Hill Member of the Favret Formation are of late Anisian age, and the overlying Panther Canyon Member of the Augusta Mountain Formation is assigned to the upper Ladinian substage. Thus, the Home Station is largely of early Ladinian age and is generally correlative with the upper member of the Prida Formation in the East and Humboldt Ranges (Fig. 5).

Panther Canyon Member. The Panther Canyon Member of the Augusta Mountain Formation is a sequence of supratidal dolomite overlain locally by clastic rocks. It is the oldest regionally extensive part of the Star Peak Group, and the dolomitic part of it maintains a constant thickness and lithology except in the west where it interfingers with more open-platform carbonate rocks. The Panther Canyon is the only part of the Star Peak Group recognized in the group's easternmost exposures in the Shoshone Mountains (Nichols, 1971). Because of the regional uniformity of the Panther Canyon, its exposures in both the eastern and central Star Peak regions are discussed together here.

Owing to localized relative uplift and erosion prior to its deposition

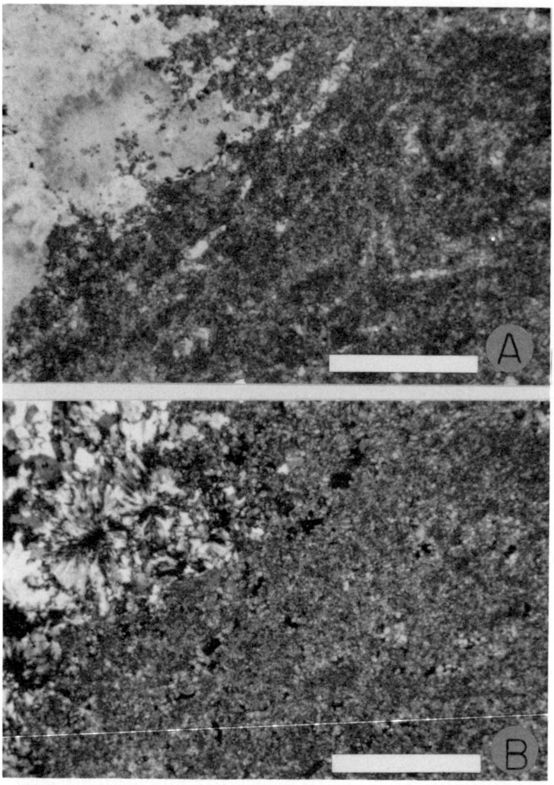

Figure 11. Photomicrographs of cherty secondary dolomite a few metres below base of the Panther Canyon member of the Augusta Mountain Formation, Favret Canyon, Augusta Mountains. A, plane light; B, crossed nicols. Fine-grained saccharoidal dolomite; patch in upper left is length-slow spherulitic chalcedony. Bar = 0.50 mm.

during late Ladinian time, the Panther Canyon Member rests in places directly on the noncarbonate rocks of the Koipato Group or Havallah sequence and elsewhere on varying thicknesses of secondary dolomite that replaces Star Peak Group carbonate rocks (Fig. 2; Nichols, 1974). Figure 12 shows the variations in total thickness of those Star Peak rocks that lie beneath the upper Ladinian and outlines the areal extent of erosion prior to deposition of the Panther Canyon. At China Mountain, the erosion surface at the base of the Panther Canyon truncates, from east to west, successively older Middle Triassic ammonite beds and thus forms a slight angular uncomformity (Nichols, 1972). The nature of the tectonism that led to this major episode of erosion is not certain, but various possibilities are discussed below. The regionally coextensive Panther Canyon supratidal dolomite and the altered rocks that underlie it indicate dolomitization by solutions genetically associated with the Panther Canyon Member or the unconformable surface upon which it was deposited (Nichols, 1974).

The Panther Canyon Member is an ususual lithologic unit. In its northeastern exposures the association of buff or yellow dolomite with overlying maroon conglomerate and mudstone produces strikingly colorful outcrops. The clastic rocks wedge out to the west and south, but the dolomite maintains a constant thickness of about 40 m throughout the region. No discontinuity is apparent between the clastic rocks, where present, and the overlying limestone of the Smelser Pass Member. Fossils are not found in the Panther Canyon, but its age, as based on stratigraphic position, appears to be late Ladinian. In the East and Humboldt Ranges, the youngest limestone beneath the Panther Canyon contains lower Ladinian fossils, and the Desatoyense Zone of early Karnian age occurs in the rocks that overlie the Panther Canyon at China Mountain (Fig. 5). The Panther Canyon Member is generally correlative in the western region with parts of both the Congress Canyon Formation and the upper member of the Prida Formation.

At its type locality in Panther Canyon near China Mountain, the Panther Canyon Member rests with angular unconformity on the China Mountain Formation of the Koipato Group. At China Mountain the member overlies

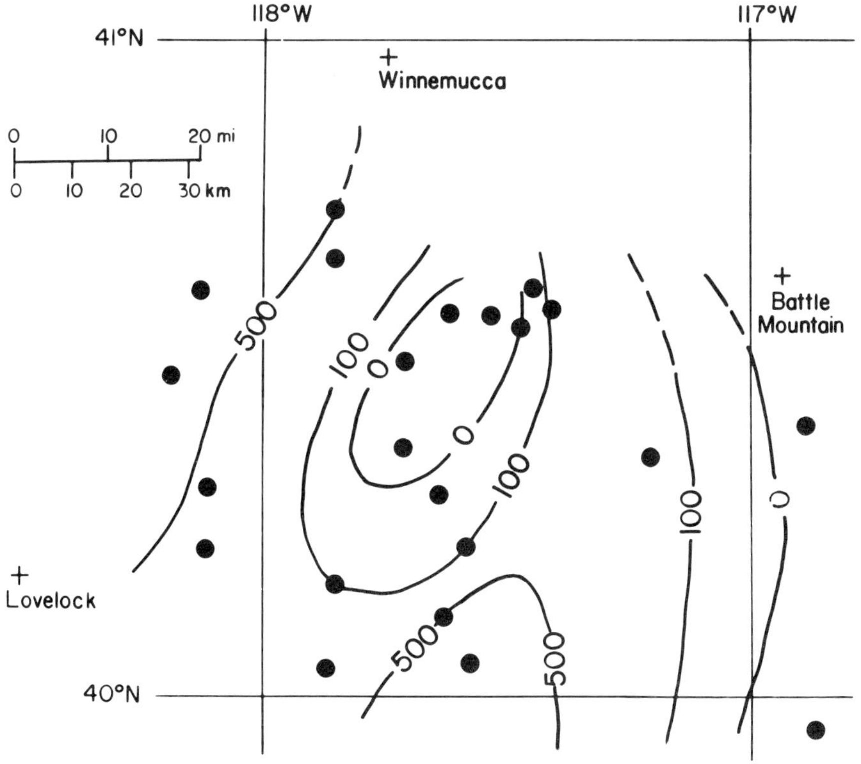

Figure 12. Isopach map, in metres, of Star Peak Group strata beneath the Panther Canyon Member, Augusta Mountain Formation, and laterally equivalent upper Ladinian rocks. Dots are control points.

massive secondary dolomite and pelmicrite of the Fossil Hill Formation, which were originally assigned by Ferguson and others (1951a) to the Favret Formation of their "upper plate facies." The Panther Canyon was also recognized originally at China Mountain and in the Fish Creek Mountains. It is now apparent that strata assignable to the Panther Canyon Member occur throughout much of the Star Peak outcrop area and that datable carbonate rocks are partially or completely absent beneath the Panther Canyon owing to uplift and erosion during Triassic time and to differential replacement of the limestone by dolomite (Nichols, 1974). Confusion between the primary dolomite of the Panther Canyon and the underlying secondary dolomite has long been one of the major obstacles to understanding the Star Peak carbonate complex. Previous workers have treated the secondary dolomite bodies as stratigraphic units rather than as alteration bodies having irregular thicknesses. The rocks included here in the Panther Canyon Member have previously been assigned to a number of different formations. As previously mentioned, they correspond to the typical middle member of the Augusta Mountain Formation (Muller and others, 1951). The Panther Canyon Member and the secondary dolomite directly underlying it form the basal part of the unit named the "Natchez Pass Formation" by Muller and others (1951) at its type locality in the northern East Range (Fig. 3B); at other places these rocks have been mapped as part of the Prida, Favret, Natchez Pass, and Augusta Mountain Formations (Ferguson and others 1951a, 1951b, 1952; Muller and others, 1951; Silberling and Roberts, 1962; Silberling and Wallace, 1969). This misidentification not only masked an important break in sedimentation but also contributed to the erroneous idea of two different Triassic facies: the "Augusta" and "Winnemucca sequences," supposedly juxtaposed by large-scale Mesozoic thrusting on the Golconda thrust. The Panther Canyon dolomite is fine grained, thickly parted, laminated, and weathers a distinctive yellowish-brown color (10YR 6/6). Outcrops have a distinctive angular, blocky appearance (Fig. 13). Fresh exposures of these rocks are tan (5Y 7/2) to gray (5YR 7/1) silty dolomicrite, and small clusters of sparry calcite or dolomite give the rocks a sparkling bird's-eye texture. Dolomite-filled mudcracks are common, and some of the surface pits are cubic in outline, suggesting that they were filled with halite. The undulant laminae (Fig. 14) resemble stromatolitic structures reported from Phanerozoic rocks by other workers. Stromatolitic algal laminations, fine-grained texture, and desiccation features identify the Panther Canyon dolomite as a penecontemporaneously dolomitized supratidal deposit (Nichols, 1974). Its deposition required a broad, stable, flat area near sea level. Therefore, this unit is essentially coeval on a regional scale and provides a useful datum, as shown in Figure 2. Photomicrographs of stratigraphically controlled samples depicting the progressive and complete dolomitization of the limestone

Figure 13. Dolomite of the Panther Canyon Member, Augusta Mountain Formation, showing its distinctive blocky outcrop character. Spur south of Favret Canyon, Augusta Mountains.

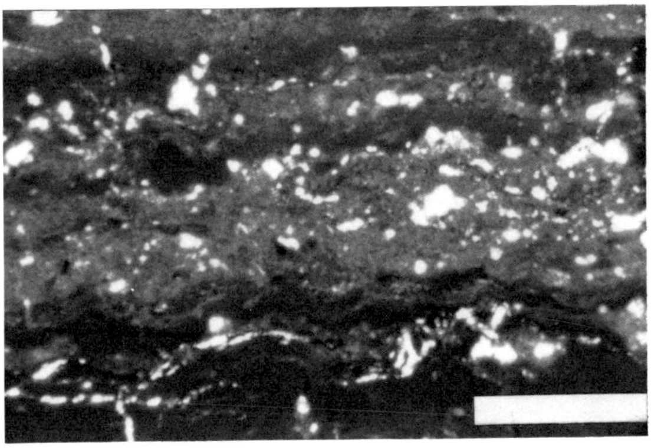

Figure 14. Photomicrograph of dolomite from the Panther Canyon Member, Augusta Mountain Formation, near Panther Canyon in the southern Sonoma Range. Irregularly laminate and stromatolitic dark cryptocrystalline dolomite incorporates bird's-eyes of clear crystalline dolomite. Bar = 0.50 mm.

beneath the Panther Canyon Member are shown in Figure 21 and in Nichols (1974). The supratidal dolomite of the member interfingers with dolomitized intertidal calcareous mudstone in the East and Stillwater Rangers (Figs. 2, 4). Farther west in the Humboldt Range, the dolomite tongues out into the intertidal and lagoonal carbonate rocks which in turn correlate with the platform-margin deposits of the Congress Canyon Formation. In the vicinity of China Mountain and the northern Fish Creek Mountains, approximately 30 m of clastic deposits rest on the Panther Canyon dolomite. The upper part of the member here is interbedded maroon-colored

(5R 3/4) mudstone, sandstone, and conglomerate and subordinate greenish-gray (10Y 7/4) mudstone. Lithologic composition, thickness, and grain size vary both vertically and horizontally. The conglomerate is massively bedded, 3 to 6 m thick, and is composed of generally well-sorted and well-rounded pebbles and cobbles of chert. Geographic variation in the particle size defines a distinct areal pattern which indicates transport from a northeasterly source, but sedimentary structures suggestive of strong unidirectional currents are absent. The finer grained rocks interbedded with the conglomerate include sandstone, mudstone, and argillite. The most spectacular clastic rocks in the Panther Canyon section are the red-colored pseudo-oolitic and pisolitic beds (Fig. 15) which are as much as 5 m thick, exhibit channeling and cross bedding, and rest on red mudstone units. The pseudo-oolites and pisolites originated as diagenetic solution structures, and their progressive development as observed in the field is described by Nichols (1972, p. 79, 82).

The association of coarse- and fine-grained redbeds in the upper part of the Panther Canyon Member indicates a terrestrial sedimentary environment. The thick beds of conglomerates most likely represent alluvial deposits that were derived from a source area to the northeast, whereas the fine-grained sediments probably originated as bog-soil or iron-pan debris (Nichols, 1972). South and west of China Mountain and the Fish Creek Mountains, the clastic rocks at the top of the Panther Canyon Member are less well represented, or absent. In the southern Tobin Range, less than 10 m of siliceous sandstone and chert pebble conglomerate mark the top of the Panther Canyon; in the East Range, northern Stillwater Range, and east-central Humboldt Range, only thin clastic interbeds and stringers of chert sand and pebbles occur in dolomite near the top of the member. In the Augusta Mountains the Panther Canyon member is composed entirely of dolomite.

Figure 15. Cross-bedded pisolitic sandstone within the clastic upper part of the Panther Canyon Member, Augusta Mountain Formation, west side of China Mountain.

Smelser Pass Member. The name "Smelser Pass" is proposed here for the limestone, capped by minor mafic volcanic rocks, that directly overlies the Panther Canyon Member in the vicinity of Smelser Pass at the north end of China Mountain (Fig. 3D). The Smelser Pass Member is the same lithologic unit as the previously unnamed upper member of the Augusta Mountain Formation of the "upper plate facies" as defined by Muller and others (1951). They originally mapped this unit as the entire Augusta Mountain Formation at China Mountain, in the Fish Creek Mountains, Augusta Mountains, and southern Sonoma Range. The Smelser Pass can also be recognized throughout the western Star Peak region, where it was previously thought to be part of the "lower plate facies" and was mapped as part of the "Natchez Pass Formation" by Ferguson and others (1951a). In the Humboldt Range it is the unit informally designated as the "lower member of the Natchez Pass Formation" by Silberling and Wallace (1969).

The Smelser Pass Member is composed mainly of limestone, but at its top the unit includes mafic volcanic and volcaniclastic rocks as much as 30 m in thickness. The carbonate rocks maintain a constant thickness of 150 to 200 m in the eastern and central parts of the Star Peak region, whereas in the Humboldt Range to the west the stratigraphic position of the Smelser Pass is occupied by the platform-margin deposits of the Congress Canyon Formation. In the southern part of the Humboldt Range, the section of Smelser Pass carbonate rocks is thickened by the interfingering of volcanic rocks (Silberling and Wallace, 1969). In the eastern and central regions, the contact between the Smelser Pass Member and the underlying Panther Canyon Member is well defined where the youngest part of the Panther Canyon consists of clastic rocks; elsewhere the contact occurs within a 3-m sequence of interbedded dolomite and limestone. The contact between the volcanic rocks of the Smelser Pass and the overlying conglomerate and carbonate rocks of the Cane Spring Formation is sharp and marks a regional episode of nondeposition and erosion during Karnian time. Evidence for this event is clear at China Mountain, where erosion has locally removed the Smelser Pass volcanic rocks beneath the conglomerate, sandstone, and deeply weathered lithified gravels of the Cane Spring Formation (Nichols, 1972).

Age-dating fossils are scarce in the Smelser Pass Member in the eastern and central Star Peak regions. Ammonites collected from the uppermost 3 m of the Smelser Pass limestone at China Mountain (Nichols, 1972) belong in part to the same species as the early Karnian fauna of the Desatoyense Zone described by Johnston (1941) from the New Pass Range.

The carbonate rocks of the Smelser Pass Member compose a monotonous succession of blocky weathering, mottled, bioturbated, medium- to thick-bedded, dark-gray (5YR 3/1), cliff-forming limestone (Fig. 16). In parts of the East and Humboldt Ranges, these rocks have been tectonically

Figure 16. Characteristic mottling and bedding of the Smelser Pass Member, Augusta Mountain Formation, in its type area on the west side of China Mountain.

broken and extensively recrystallized by metamorphism. Chert blebs about 10 cm in diameter are common, as are large solution collapse features. Bedding surfaces are characteristically lumpy.

Petrographically, the Smelser Pass limestones are partly to completely altered, clotted, pelleted carbonate mudstones and minor amounts of skeletal, intraclastic, or oolitic grainstone (Figs. 17A to 17E; Fig. 18A). The ooids are well sorted, and many of them are outlined by micritic envelopes as are the poorly sorted skeletal fragments (Figs. 17C, 17D). The ooids have been leached and subsequently infilled with a mosaic of drusy calcite (Figs. 17A, 17B).

The depositional environment of the Smelser Pass Member carbonate rocks appears to range from open-platform intertidal to shallow subtidal, as the rocks are analogous to the skeletal mud facies of the Bahamas described by Purdy (1963). The ooids were most likely washed into the mud facies from a contiguous oolite shoal (Fig. 18C).

The mafic volcanic rocks in the upper part of the Smelser Pass Member comprise a heterogeneous assemblage of dense or amygdaloidal, nonporphyritic, massive flows, as well as breccia, tuff, and tuffaceous argillite. In the China Mountain area, these volcanic rocks are limited to small pods that fill depressions in the underlying limestone. They lack mappable thickness in the Fish Creek and Augusta Mountains, whereas in the East, Stillwater, and southern Tobin Ranges they form a continuous unit a few tens of metres thick at the top of the Smelser Pass Member. These volcanic rocks are poorly represented in the northern Humboldt Range, as described

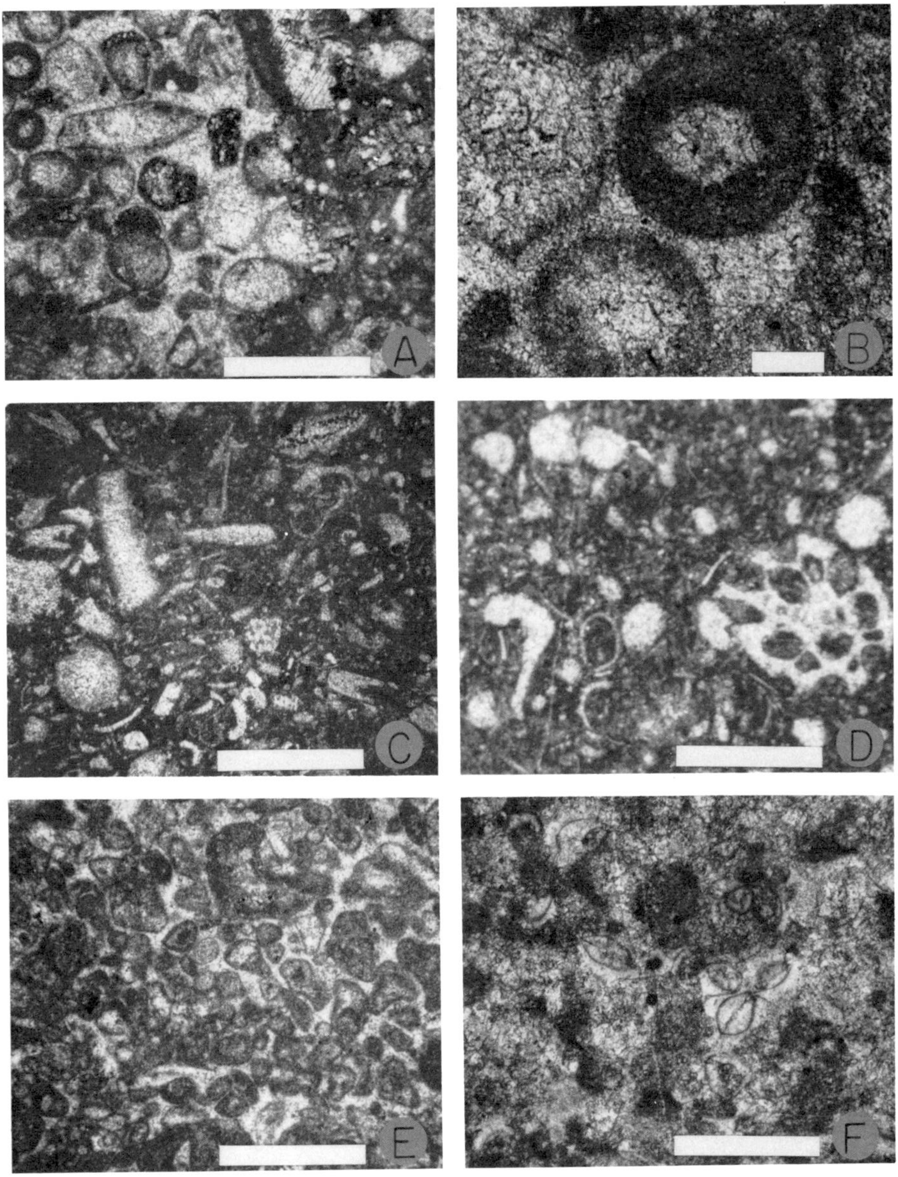

Figure 17. Photomicrographs of the Smelser Pass Member, Augusta Mountain Formation (A-E), and the Cane Spring Formation (F). All photomicrographs except D are from China Mountain, northern Tobin Range; D is from the southern Tobin Range. A, intraclastic oosparite showing "ghosted" intraclasts. Recrystallized calcite matrix transects boundaries of some of the pelmicrite intraclasts. Note echinoderm plate at upper right. Bar = 0.50 mm. B, magnification of part of Figure 17A. Ooids in various stages of diagenetic alteration to sparry calcite. Bar = 0.10 mm. C, biomicrite. Note fragment of calcareous algae at lower right of center. Bar = 0.50 mm. D, biomicrite. Possible ooids are completely altered to sparry calcite. Bar = 0.50 mm. E, fossiliferous oosparite having partly altered micritic intraclasts. Bar = 0.50 mm. F, fossiliferous intrasparite having relict pelmicrite patches and conspicuous ostracod shells. Bar = 0.50 mm.

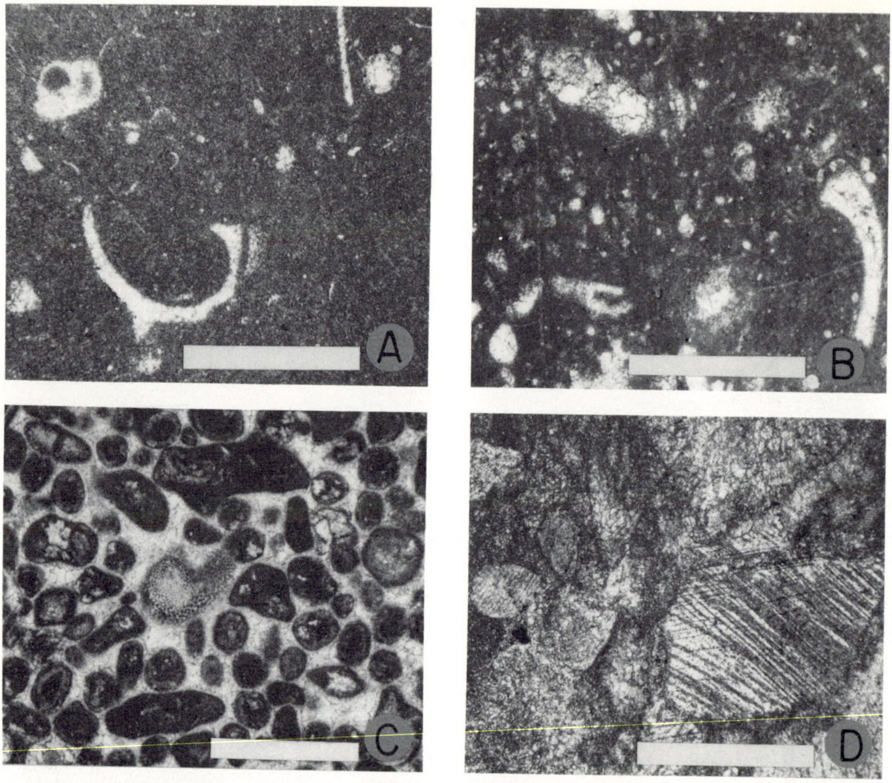

Figure 18. Photomicrographs of the Smelser Pass Member, Augusta Mountain Formation (A, B, and C), and the Congress Canyon Formation (D). A is from China Mountain; B is from the northern Stillwater Range; C is from American Canyon, southern Humboldt Range; and D is from Congress Canyon in the northern Humboldt Range. A, B, biomicrite. Bar = 1.00 mm. C, intraoosparite. Some grains are partly replaced by clear dolomite. Note echinoderm grain at center. Bar = 0.50 mm. D, altered coarse-grained limestone. Bar = 1.00 mm.

by Silberling and Wallace (1969), possibly because of erosion prior to deposition of the Cane Spring Formation. In the central Humboldt Range, the mafic volcanic rocks appear as a continuous unit 10 to 30 m thick at the top of the Smelser Pass. Farther south in the Humboldt Range they thicken abruptly by interfingering with the Smelser Pass, forming at least half of the member's abnormally great thickness here of about 400 m.

Cane Spring Formation

The Cane Spring Formation was named by Muller and others (1951) for a thick sequence of carbonate rocks that overlies the Augusta Mountain Formation on the summit of the Augusta Mountains. On the southeast flank of the Augusta Mountains, it forms a dip slope just beyond the Sonoma Range quadrangle. They estimated its thickness as "more than 1,000 feet," but the type area has never been mapped in detail, and the top of the formation is not exposed here. Muller and others (1951) described a "thin lava flow" as the base of the type section, but this apparently is part of the mafic volcanic unit that locally forms the top of the Smelser Pass Member of the Augusta Mountain Formation. At the summit of the Augusta Mountains, these volcanic rocks are directly overlain by several metres of silliceous pebble conglomerate that is evidently the "20 foot lens of . . . conglomerate [originally described as being] 100 feet above [the] base" of the Cane Spring Formation (Muller and others, 1951). To our knowledge, nowhere do carbonate rocks separate the mafic volcanic rocks from the clastic rocks of the Cane Spring. Instead, the basal clastic strata of the Cane Spring evidently rest upon a widespread unconformity.

Siliceous clastic rocks characterize the base of the Cane Spring Formation only in its more eastern and northern exposures; in the southwestern part of the Star Peak outcrop region, the basal part of the formation consists of impure, brown-weathering, commonly nodular and evenly bedded, shelly limestone, like that which overlies the basal clastic rocks elsewhere. Calcareous basal beds of this kind rest on the Smelser Pass volcanic rocks at one place in the core of an open anticlinal fold in the exposures of the Cane Spring along the west flank of the Augusta Mountains. Hence, not only the top but also the base of the formation is exposed here, although the basal contact is not shown on the small-scale map by Muller and others (1951). Unfortunately, the faults that cut this exposure have not been mapped, so the thickness and stratigraphic succession of the Cane Spring here are as yet undetermined. However, the Cane Spring Formation is similar in character in the nearby southern Tobin Range where, based on detailed mapping, Burke (1973) determined its thickness to be about 400 m.

Where present in the eastern and central Star Peak regions, the Cane Spring basal clastic rocks consist of poorly bedded, massive, dusky red (5R 3/4) to brown-gray (5YR 4/1) weathering, siliceous conglomerate and subordinate interstratified units of quartzitic sandstone. In the Augusta and Fish Creek Mountains, where the maximum thickness of basal clastic strata is about 10 m, conglomerate clasts are mainly pebble size and smaller. Farther north at China Mountain, however, the basal clastic unit is as much as 50 m thick and includes cobble-size clasts in massive, laterally persistent units which are separated by planar cross-bedded reddish-brown (10R 7/4) quartzitic sandstone. Sporadic lenses as much as 5 m thick within the conglomerate and sandstone consist of pasty looking, grayish-orange clastic rocks in which original sedimentary fabrics and structures have been partly to largely obliterated (Nichols, 1972, Pl. llx). These are interpreted by Nichols (1972) as recording episodes of nondeposition, deep weathering, and soil development that occurred during formation of the Cane Spring basal clastic rocks. Irregular cylindrical structures, as much as 10 cm in diameter and perpendicular to the bedding, are thought to be casts of tree roots or trunks.

At China Mountain the basal conglomerate of the Cane Spring Formation fills channels eroded into the top of the underlying Smelser Pass Member of the Augusta Mountain Formation. In addition, subaerial exposure that occurred between deposition of the Smelser Pass and Cane Spring strata resulted in the formation of caves, some of which extend tens of metres down into the Smelser Pass limestone. Some of these caves are filled with Cane Spring conglomerate (Nichols, 1972). Farther south and west, where calcareous rocks are at the base of the Cane Spring, disconformity is not so evident. However, the abrupt upward disappearance of the Smelser Pass volcanic rocks and the widespread and abrupt vertical change in the nature of sedimentary character suggest disconformity. The few occurrences of datable fossils in the Smelser Pass Member and Cane Spring Formation (Fig. 5) indicate that a stratigraphic hiatus may separate the two units throughout the outcrop region.

The medium- to thick-bedded brownish-weathering limestone that overlies the basal clastic rocks, or where these are absent, forms the base of the Cane Spring Formation, grades upward within several tens of metres into more thickly bedded gray limestone which makes up most of the section. On the outcrop only subtle differences in color, skeletal content, mottling, and bedding characteristics can be distinguished among these rocks that tend to be monotonously alike. They are more interesting petrographically (Figs. 17F, 19).

Most of the Cane Spring limestone is mud-supported, more or less pelleted wackestone, generally with abundant skeletal allochems. Oolitic or other grains suggestive of proximity to the platform margin, such as the ooids found widely scattered within the Smelser Pass wackestone, have not been observed. Variation in the size of molluscan skeletal grains is the most conspicuous difference among the beds. Thick, massive beds tend to have abundant large fragments of bivalve and gastropod shells; nearly unbroken gastropod shells having heights as great as 15 cm in cross section are not uncommon in the Augusta Mountains. Sporadic higher energy interbeds of grainstone (Figs. 19C, 19D) mainly contain thoroughly micritized, well-rounded shell fragments. Petrographic evidence for near-surface diagenesis influenced by meteoric water is abundant where original fabrics are well preserved. Molluscan shell fragments that were evidently originally aragonitic in composition are represented by sparry calcite mosaics that fill original void space which resulted from diagenetic dissolution of all but the micritic envelopes outlining the original grains. However, other

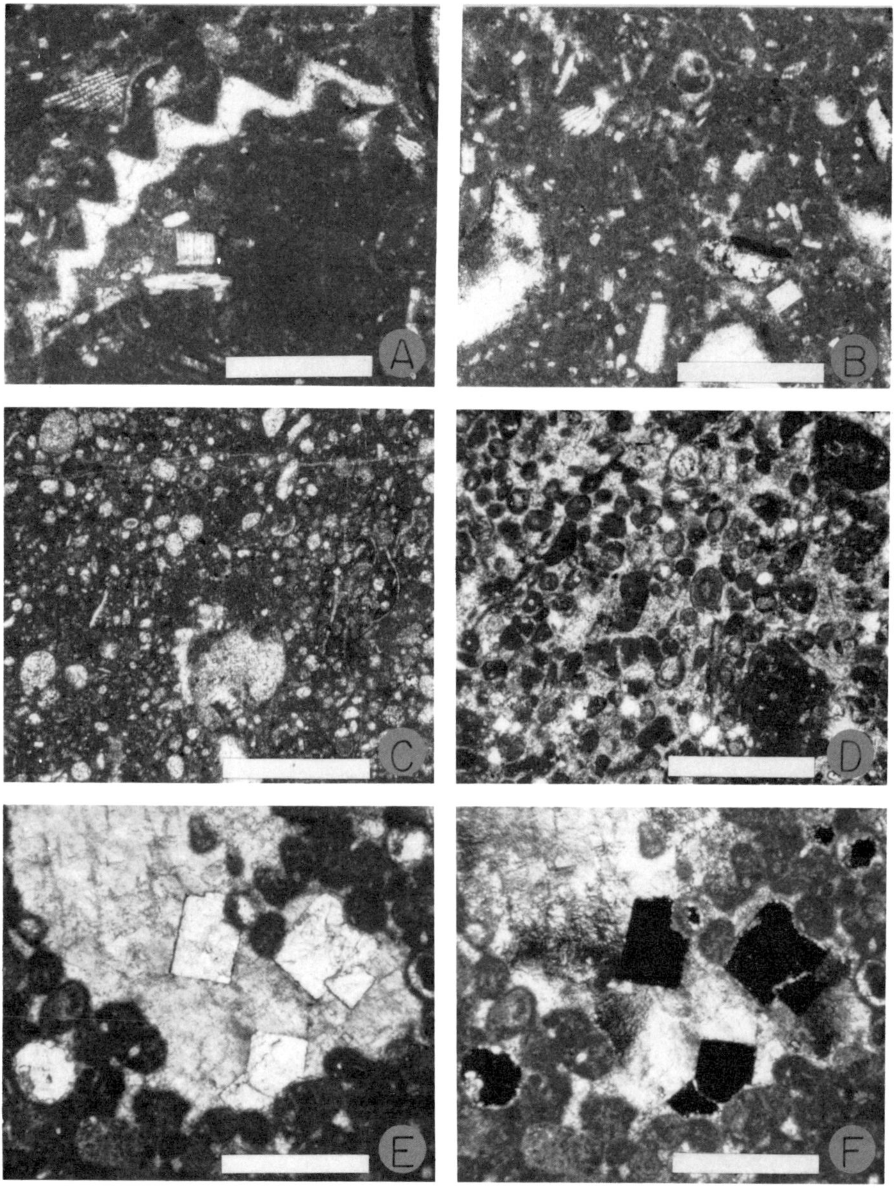

Figure 19. Photomicrographs of the Cane Spring Formation from near Jenkins Ranch, Augusta Mountains. A, B, molluscan-echinoderm wackestone. Some of the larger shell fragments, such as the plicate bivalve shell fragment in A, are mosaics of sparry calcite suggestive of void fillings following dissolution of original aragonite. Rectangular grain at lower left-center of A retains original prismatic shell structure. Lath-shaped prisms and aggregates of prisms are scattered through B. Bar = 0.50 mm. C, neomorphic biomicrite. Light grains are neomorphic spar representing well-rounded molluscan(?) fragments. Bar = 0.50 mm. D, biopelsparite. Bar = 0.50 mm. E, F, (E, plane light; F, crossed nicols), Pelletoid micrite. Original evaporite mineral(?) replaced by large cloudy calcite crystals and isotropic cubic crystals of fluorite.

molluscan grains, commonly visible in the same thin sections (Fig. 19A), preserve their original prismatic shell structure. These shells were apparently calcitic in composition and appear to demonstrate the differential diagenetic solution of aragonite such as that observed in many Pleistocene limestones whose diagenesis has been influenced by meteoric waters. The diagenetic, textural, and paleontologic characteristics of the Cane Spring limestone indicate intertidal and lagoonal, inner-platform deposition, which was interrupted perhaps occasionally by subaerial exposure.

Thin sections of wackestone from near the top of the Cane Spring section on the west flank of the Augusta Mountains contain coarsely crystalline patches up to several millimetres across. These resemble ''bird's-eyes'' but are filled with large cloudy anhedral crystals of calcite (Figs. 19E, 19F) rather than sparry cement. This texture suggests calcite replacement of original sulfate nodules; however, cubic crystals included in these patches are composed of fluorite which may indicate hydrothermal rather than evaporitic diagenetic origin of these structures.

Age-diagnostic fossils are unknown from the Cane Spring Formation in the eastern and central Star Peak regions. However, calcareous beds near the base of the overlying Osobb Formation in the Augusta Mountains contain tibetitid, clionititinid, and juvavitinid ammonites. None of these belong to described species, but they have counterparts in better known faunas from the Kerri Zone. Thus, the upper limit of the Cane Spring Formation, and hence of the Star Peak Group, in at least this part of its outcrop region, coincides approximately with the Karnian-Norian inter-stage boundary.

CENTRAL REGION

The central region encompasses the southern Sonoma Range, China Mountain, east-central East Range, and the area that includes the more northerly of the Star Peak exposures in the southern Tobin Range. Deposition of the Star Peak Group in the central outcrop region was greatly influenced by two episodes of uplift and erosion during Middle Triassic time. Strata older than late Anisian either were not deposited here or were eroded away during early Anisian uplift, which is described in the discussion of the lower member of the Prida Formation in the western part of the Star Peak region. In the central region, strata assignable to the lower member of the Prida are too thin to be recognized as a separate unit, so that the rock unit that represents the Fossil Hill Member of the Prida Formation farther west rests in the central region directly on the Koipato Group or, where the Koipato was removed by erosion, on the Golconda allochthon. As the upper member of the Prida does not extend this far east, the Fossil Hill stands alone and is therefore of formational rank in the central part of the region (Fig. 5). These rocks have been assigned previously to the Prida Formation (Ferguson and others, 1951a; Muller and others, 1951), the "middle member" of the Prida Formation (Nichols, 1974), and to the Favret Formation (Ferguson and others, 1952).

Owing to uplift and erosion during Ladinian time, some or all of the Fossil Hill Formation in the central part of the Star Peak region was eroded away prior to deposition of the Panther Canyon Member of the Augusta Mountain Formation. Where at least partly preserved, some or all of the Fossil Hill limestone immediately below the Panther Canyon has been completely altered to dolomite. Thus, in the central region the basal part of the Star Peak Group may be the Fossil Hill Formation, secondary dolomite formed from the Fossil Hill, or the Panther Canyon Member of the Augusta Mountain Formation.

Fossil Hill Formation

In the central region the Fossil Hill Formation is essentially like the corresponding part of the section to the west, where the unit forms the middle member of the Prida Formation, and to the southeast, where it forms the upper member and major part of the Favret Formation. In comparison with the Fossil Hill farther southeast, the formation in the central region is less shaly. Thin- to medium-thick beds of dark, slightly quartz-silty fossiliferous lime mudstone predominate and are intercalated with platy weathering calcareous siltstone and shale (Fig. 20). The northward increase in the relative proportion of bedded limestone within the member is, however, too gradational to be mapped. Therefore, all of these rocks are included in the same lithologic unit. Ammonoids and Daonellas from near the base of the Fossil Hill in the central region represent the Rotelliformis Zone. Thus, the base of the section here is much younger than the base of the Fossil Hill Member of the Favret Formation, but it does correlate with the base of the Fossil Hill farther west in areas affected by early Anisian uplift.

On the northeast side of China Mountain, as much as 100 m of Fossil Hill limestone is preserved beneath the secondary dolomite, below the Panther Canyon. The Fossil Hill limestone is progressively lost in a southwesterly direction, and it is absent where the Star Peak crops out in the southwestern part of China Mountain and the southernmost part of the Sonoma Range. In addition, successively older ammonite faunas of the Meeki and Rotelliformis Zones within the Fossil Hill Formation are truncated by the sub-Panther Canyon dolomite in a southwesterly direction (Nichols, 1972, Fig. 10) at China Mountain. Even though the thickness of the secondary dolomite varies greatly up to as much as 75 m, the thickness of the limestone involved and the distance over which

Figure 20. Thin- to medium-bedded limestone intercalated with platy weathering, fine-grained calcareous clastic rocks. Fossil Hill Formation, southern Tobin Range.

truncation takes place are great enough so that this pattern is necessarily the result of tilting and erosion and secondary dolomitization prior to Panther Canyon deposition.

A similar progressive truncation of Fossil Hill calcareous strata exists beneath the sub-Panther Canyon secondary dolomite in the southern Tobin Range, north of the area where the Fossil Hill forms part of the Favret Formation. In the most southern sections where the Fossil Hill Formation is wholly late Anisian in age, as much as several tens of metres of limestone is preserved beneath the secondary dolomite. These calcareous rocks pinch out northward within the southern Tobin Range. Beyond their northern limit, the basal part of the Star Peak Group consists of 20 to 30 m of sugary crystalline dolomite that underlies, with a sharp upper contact, the fine-grained "primary" dolomite of the Panther Canyon Member of the Augusta Mountain Formation. By extrapolation, this basal secondary dolomite was originally the lower uneroded part of the Fossil Hill Formation. The lowest 5 to 10 m of the crystalline dolomite is in places sandy and impure; it may represent the original impure carbonate rocks—akin to the lower member of the Prida Formation farther west—at the base of the Fossil Hill Formation.

The Fossil Hill Formation therefore thins by truncation and dolomitization of its upper part in a southwesterly direction from China Mountain and in a northwesterly direction from the southern Tobin Range, toward an area along the west-central East Range and the southwesternmost part of the Sonoma Range. This area underwent the greatest relative uplift and erosion during Ladinian time, and here the Panther Canyon Member rests directly on the Koipato Group.

The Fossil Hill Formation at China Mountain is mainly thin- to medium-bedded, finely crystalline, dark-gray (5YR 3/1) limestone interbedded with calcareous shale and siltstone (Nichols, 1972, 1974). Its upper part has been altered to massive saccharoidal, light-gray (5YR 7/1) dolomite which attains a maximum thickness locally of 75 m. A transition zone of mottled limestone, 3 to 6 m thick in which the mottles consist of dolomite, everywhere separates the unaltered limestone from the overlying secondary dolomite. The unaltered limestone and saccharoidal dolomite are easy to distinguish from one another in the field, as the former is slope forming in contrast to the massive, benchlike outcrops of the latter.

Thin-section samples (Fig. 21) show the progressive dolomitization upsection of stratigraphically controlled samples of Fossil Hill limestone. The unaltered Fossil Hill is a slightly quartz-silty, more or less fossiliferous intrapelmicrite or pelmicrite (Fig. 21A). The intraclasts are composed of either pelmicrite or biomicrite. Rocks transitional between dolomite and limestone are characterized by dense pellets and partially dolomitized intraclasts in a groundmass of subhedral to euhedral dolomite. The end result is a mosaic of euhedral dolomite (Fig. 21C).

Augusta Mountain and Cane Spring Formations

The parts of the Augusta Mountain and Cane Spring Formations that occur in the central region are essentially like those in the eastern region; they have been described under that heading. In the central, as in the western region, the Home Station Member of the Augusta Mountain Formation is absent. Thus, the Panther Canyon Member in the central region rests directly on secondary dolomite that formed by diagenetic alteration of the Fossil Hill Formation, or on rocks older than the Star Peak Group.

WESTERN REGION

The formations characteristic of the western part of the Star Peak area are exposed in the northern East Range, Humboldt Range, and northern Stillwater Range. At these places the Prida Formation makes up the lower part of the Star Peak Group and thickens to the northwest, where it forms the major part of the group. Where the Prida is thickest, its upper part is basinal limestone which is overlain by, and presumably also interfingers laterally with, platform-margin carbonate deposits of the Congress Canyon

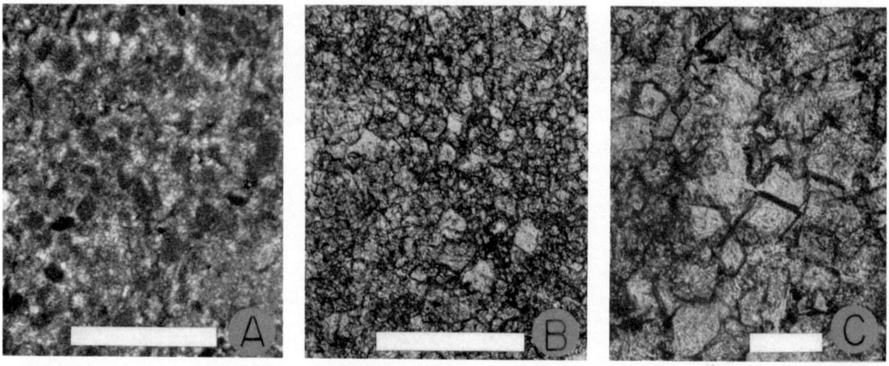

Figure 21. Photomicrographs of the Fossil Hill Formation from China Mountain in the northern Tobin Range. A, incompletely altered silty, pelletal limestone. Micritic matrix has been largely replaced by microcrystalline dolomite which has started to invade the pellets. Bar = 0.50 mm. B, inequigranular dolomite of xenotopic to hypidiotopic fabric which completely replaces original pelmicrite matrix. Bar = 0.50 mm. C, magnification of Figure 21B showing euhedral dolomite rhombs. Bar = 0.10 mm.

Formation (Figs. 2, 4). Farther east, strata that are assignable to the Prida Formation are overlain and laterally replaced by secondary dolomite of various thicknesses. Successively above these units are the platform deposits of the Panther Canyon and Smelser Pass Members of the Augusta Mountain Formation which are more inner platform equivalents of the Congress Canyon. The Cane Spring Formation disconformably overlies these platform carbonate units and forms the upper part of the Star Peak Group, as it does everywhere.

Prida Formation

Three lithologic units of approximately equal thickness were originally described by Muller and others (1951) at the type locality of the Prida Formation in the northern East Range, where the entire formation is commonly about 400 m thick. Subsequently, these three units were recognized in the Humboldt Range where they were established as informal members of the Prida Formation by Silberling and Wallace (1967, 1969), although locally they are thicker here than in the type area.

The lower member of the Prida Formation is variable lithologically and in thickness. This diversity reflects both its position overlying surface relief on the eroded Koipato Group and local tectonic activity during and after its deposition. The lower member is as much as 120 m thick and includes, in order of decreasing abundance, locally derived terrigenous clastic rocks, secondary dolomite, limestone of various kinds, and mafic volcanic rocks. The middle member of the Prida Formation is here named by us the Fossil

Hill Member. It ranges up to more than 100 m in thickness and consists of calcareous siltstone, mudstone, and shale interbedded in various proportions with dark micritic limestone containing exclusively pelagic ammonite and bivalve shells. The upper member of the Prida Formation is dark, commonly laminated and cherty limestone that is the basinal equivalent of all the carbonate-platform deposits of the Congress Canyon and Augusta Mountain Formations. As these regressively prograding platform deposits thin to the northwest, the upper member thickens to a maximum of several hundred metres in the northern Humboldt Range.

Deposition of the more northwestern and basinward parts of the upper member of the Prida Formation was not interrupted by the Ladinian episode of uplift and erosion that affected the central Star Peak region. However, the older parts of the upper member, the subjacent Fossil Hill Member, and farther on all of the lower member are unconformably truncated progressively eastward toward the uplifted area (Fig. 2). Moreover, the truncated limestone of the Prida Formation has been recrystallized to sugary dolomite to depths as much as 100 m beneath the Panther Canyon Member.

Lower Member. Abrupt lateral variations in thickness and composition of the lower member of the Prida Formation in the Humboldt Range were attributed by Silberling and Wallace (1969, p. 10–12, 25–28, Pl. 1) to substantial topographic relief on the eroded surface of the underlying Koipato Group. This explanation required that deposition of the lower member commenced as early as late Spathian time in places away from highs on the Koipato basement in the northern part of the Humboldt Range, but not until as late as late Anisian time, perhaps 5 to 10 m.y. later, atop the highs farther south and east in the range. Our recent discovery of a late Spathian *Prohungarites-Subcolumbites* ammonite fauna in the lower member of the Prida, about 1.5 km south of Fossil Hill (the location of which is shown on the geologic map by Wallace and others, 1969) in the southern Humboldt Range, now shows their interpretation to be incorrect.

Three laterally persistent units were recognized by Silberling and Wallace (1969, Pl. 1) within the lower member at places in the northern Humboldt Range where the member is as much as 120 m thick. The basal clastic unit, no thicker than a few tens of metres, is terrigenous clastic rock derived from the underlying Koipato Group. Above this is a carbonate unit, and still higher is a distinctive unit of slabby weathering homogeneous brown calcareous sandstone at the top of the member. The sandstone varies in thickness to a maximum of 100 m, largely by interfingering with the overlying limestone of the Fossil Hill Member of the Prida Formation. Limestone of the intermediate carbonate unit is generally unfossiliferous and contains much terrigenous clastic and coarse skeletal sediment. Upper Spathian ammonites, representing at least two stratigraphic levels within the *Prohungarites-Subcolumbites* beds and the uppermost Spathian Haugi

Zone, have been obtained from scattered localities (Silberling and Wallace, 1969, Pl. 1, Table 1).

Much thinner sections of the lower member, especially in the southern part of the Humboldt Range, as in the vicinity of Fossil Hill, consist of basal terrigenous clastic strata of variable thickness overlain by finely crystalline, sugary secondary dolomite which characteristically is sandy or gritty in the uppermost part beneath the Fossil Hill limestone. As much as 10 m of impure limestone and calcareous sandstone is preserved locally between the basal clastic rocks and superjacent secondary dolomite. The newly discovered late Spathian ammonites occur in this impure limestone. Our interpretation, as illustrated in Figure 22, is that the secondary dolomite of the lower member of the Prida is separated from the overlying upper Anisian beds of the Fossil Hill Member by a disconformity representing an appreciable stratigraphic hiatus. Moreover, the carbonate rocks and the small amount of associated mafic volcanic rock of the lower member in the southern and central Humboldt Range are equivalent to the carbonate unit of the lower member farther north in the range. Partial or complete dolomitization of the carbonate unit took place beneath this disconformity which locally separates the overlying brown calcareous sandstone unit from eroded remnants of the carbonate unit in the central part of the range. The brown calcareous sandstone unit and possibly the metre or so of sandy or gritty dolomite at the top of the lower member farther south are essentially the basal clastic beds of the Fossil Hill Member. These rocks are retained in the lower member of the Prida for the sake of

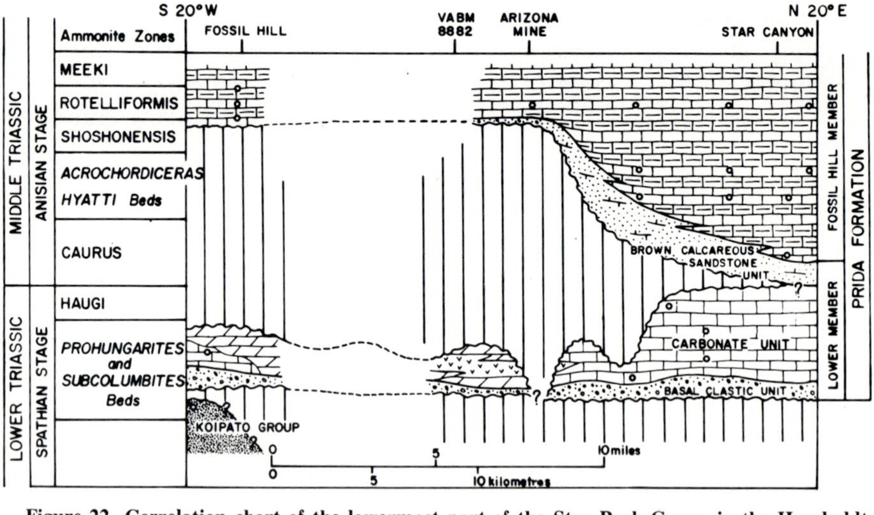

Figure 22. Correlation chart of the lowermost part of the Star Peak Group in the Humboldt Range. Small circles indicate the occurrence of age-diagnostic fossils; vertical ruling, stratigraphic hiatus.

consistency with past usage and cartographic utility.

Variations in the thickness and composition of the lower member of the Prida Formation in the Humboldt Range resulted mainly from local uplift, erosion, and secondary dolomitization during early Anisian time. These events have regional significance because the lower member of the Prida in the northern Stillwater Range is essentially like that in the southern Humboldt Range and may also represent eroded and partly dolomitized late Spathian strata. Also, in the northern East Range the thin and generally dolomitic or fine-grained clastic lower member of the Prida increases abruptly in thickness to about 200 m of calcareous sandstone, conglomerate, and pebbly sandstone in the southwesternmost of these exposures. This again suggests local tectonism during the early phase of Star Peak deposition.

Because of this tectonism and erosion only remnants of the Spathian rocks are preserved in the lower member of the Prida Formation. For the most part, however, these rocks are evidently correlative in age and originally were similar lithologically to the oldest rocks of the Star Peak Group farther to the southeast, in the Tobin Formation of the southern Tobin Range and Augusta Mountains. This similarity suggests that such strata might have blanketed a large part of the Star Peak outcrop area prior to localized uplift and erosion, which took place first during early Anisian time and again during Ladinian time.

Fossil Hill Member. The name "Fossil Hill" is proposed here for the "middle member" of the Prida Formation as defined by Silberling and Wallace (1969) from geologic mapping in the Humboldt Range. The type section of this member is on the south side of Fossil Hill in the southeastern Humboldt Range (Wallace and others, 1969b), where it attains a thickness of 60 m. Of the three members of the Prida Formation, the Fossil Hill is named formally because it extends beyond the area in which the Prida is recognized. The Fossil Hill Member has formational rank at China Mountain and in part of the southern Tobin Range where rocks resembling the lower member of the Prida are too thin to be mapped separately from the Fossil Hill and where the upper member of the Prida is absent. Elsewhere in the southern Tobin Range and in the Augusta and Fish Creek Mountains, the Fossil Hill makes up most of the Favret Formation and is regarded as a member of the Favret.

The thickness of the Fossil Hill Member in the western Star Peak outcrop region is generally about 60 m, but locally, as in parts of the northern Humboldt Range, the thickness increases to about twice this amount because older strata are included at the base (Silberling and Wallace, 1969, Pl. 2). In the northwesternmost part of the Stillwater Range, at China Mountain, and in part of the southern Tobin Range the Fossil Hill is unusually thin and locally absent because of erosion and secondary dolomitization beneath the overlying Panther Canyon member (Fig. 2). The member thickens and attains its greatest thickness of about 200 m in the Augusta Mountains, in the eastern part of the Star Peak region where the Fossil Hill forms part of the Favret Formation.

The age and correlation of the Fossil Hill Member is relatively well known owing to its wealth of cephalopods and halobiid bivalves preserved in it. Wherever the uppermost beds of the Fossil Hill are preserved, they contain fossils of the Occidentalis Zone. Thus, the upper limit of the Fossil Hill coincides approximately with the Anisian-Ladinian interstage boundary as defined by Silberling and Tozer (1968). The member's lower boundary, however, is quite variable in age, being older both east and west of the intra-Star Peak high. In the western part of the region, the base of the Fossil Hill is locally as old as the Caurus Zone, the lowermost zone of the Anisian Stage (Silberling and Wallace, 1969, p. 18). Elsewhere, in the central Star Peak region, where the Fossil Hill was affected by early Anisian uplift, fossils of the upper Anisian Rotelliformis Zone occur at or near the base of the member (Fig. 5).

Lithologically the Fossil Hill Member is fairly homogeneous and distinct from all other parts of the Star Peak Group. It consists largely of laterally discontinuous thin- or medium-thick beds of dense gray-weathering limestone mixed with a wide variety of platy weathering calcareous mudstone, calcareous siltstone, and silty limestone which typically weathers yellowish brown and pinkish gray. Figure 23 illustrates the variety of rock types exposed at Fossil Hill. Sporadic grains of quartz silt are characteristic in the dense gray limestone beds which in parts of the section become noticeably silty or even sandy and brown weathering. Shell fragments and more complete specimens of cephalopods and halobiid bivalves are abundant and in places form the main constituent of the limestone beds (Fig. 24). Flattened impressions of these kinds of fossils are also common on the parting surfaces of the fine-grained calcareous clastic rocks. Skeletal carbonate grains derived from benthonic marine invertebrates are virtually absent. In thin section (Fig. 9) the limestone beds are either clotted, pelletoidal, slightly quartz-silty wackestone or current-sorted grainstone composed of finely comminuted thin-shelled skeletal debris; some individual beds contain layers of both types of limestone.

In the Augusta Mountains, where the Fossil Hill forms the major part of the Favret Formation, dark-gray to black shaly limestone and calcareous shale are more important constituents than they are in the typical exposures in the Humboldt Range. The limestone interbeds in the Augusta Mountains tend to be concretionary, especially in the lower part of the section. The northeasternmost exposures at the northeast end of the Fish Creek Mountains contain a higher proportion of bedded dense limestone than is typical in the Fossil Hill. These lateral variations, however, are subtle and intergrade with more typical Fossil Hill rock types, so that the Fossil Hill throughout

Figure 23. Ammonite-bearing lenticular limestone beds in subfissile silty calcareous mudstone of the Fossil Hill Member, Prida Formation, in gully exposure at Fossil Hill, Humboldt Range. Limestone float debris dominates surrounding ground surface. Round holes in outcrops resulted from coring for paleomagnetism studies.

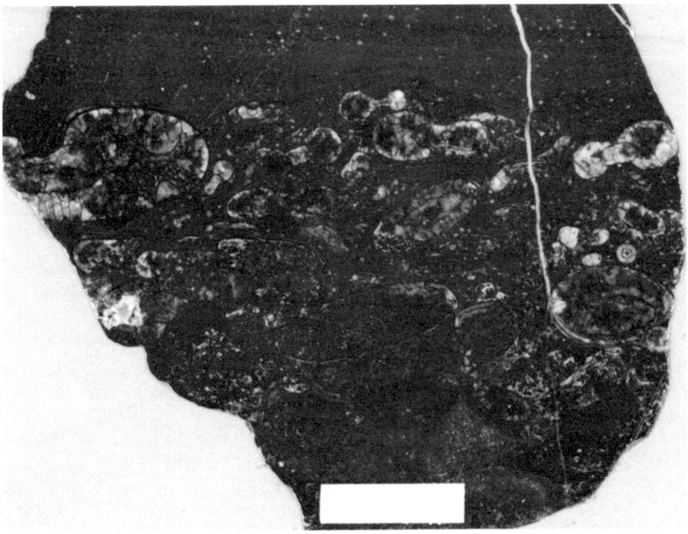

Figure 24. Polished section of ammonite-rich limestone bed from the Fossil Hill Member, Prida Formation, at Fossil Hill. Note imbrication of shells suggesting a current direction toward the right. Bar scale = 3.00 mm.

Figure 25. Photomicrographs of the upper member of the Prida Formation. Photomicrographs A and B are from the northern Humboldt Range; C is from Fisher Canyon, southern Humboldt Range; D is from the northern East Range. A, B, chert layer in partly silicified and dolomitized pelmicrite (A, plane light; B, crossed nicols). Bedding slants downward to the right. Light rectillinear area in A is fine-grained anhedral quartz pseudomorphous after original evaporite(?) mineral. Coarse crystalline calcite at right side also may be replacement of evaporite crystals; fine-grained matrix is anhedral quartz partly replaced by abundant small dolomite rhombs. Recrystallized and partly replaced pelmicrite at upper right and lower left. Late-stage calcite vein, trending diagonally upward and to the right, is about perpendicular to stratification. Bar = 0.50 mm. C, laminated calcareous mudstone having sparse biogenic allochems and quartz silt; diffuse circular sparry patches may be calcite replacements of radiolaria. Bar = 0.50 mm. D, calcareous mudstone, similar to C. Bar = 0.50 mm.

the Star Peak ourcrop area is a lithologically distinct unit.

In addition to our photomicrographs (Figs. 9, 21A) of Fossil Hill limestone from various localities, thin sections of fossiliferous wackestones containing small but complete shells of ammonoids and ostracods have been illustrated by Horowitz and Potter (1971, Pls. 44, 45) from the type locality of the member at Fossil Hill.

Although the relative duration of the Triassic ammonite zones listed in Figure 5 is unknown, the Fossil Hill Member includes considerably more zones in proportion to its thickness than do other parts of the Star Peak Group, and it thus appears to be a relatively condensed deposit.

Upper Member. The upper member of the Prida Formation is the principal expression of basinal carbonate rocks within the Star Peak Group. It is found only in the western part of the region, in the Humboldt and East Ranges, and in the small southeasternmost exposures of the stillwater Range near Fencemaker Pass.

The upper member is a monotonous sequence of dark-gray limestone which, for the most part, is recrystallized, sparsely quartz-silty, lime wackestone having a small allochem-matrix ratio (Figs. 25C, 25D). It is regularly parted into beds of medium thickness which commonly show internal lamination (Fig. 26). Dark chert is characteristically present in discontinuous laminae or thin beds and in places retains traces of original biogenic siliceous grains (Silberling and Wallace, 1969, p. 14). Subordinate interstratified units of thick-bedded, coarse-grained, massive gray limestone up to several metres thick are interpreted as slope deposits of carbonate debris from the platform margin. However, because of hydrothermal recrystallization in the Humboldt Range and northern East Range, sedimentary structures and textures that would confirm this have not been observed. More indicative of slope deposition is the sporadic occurrence, within the otherwise evenly stratified section, of slump folds and sedimentary breccia (Fig. 27), some of which contain blocks apparently derived from the carbonate platform. In keeping with the interpreted off-platform depositional environment for the upper Prida, shelled benthonic invertebrate fossils

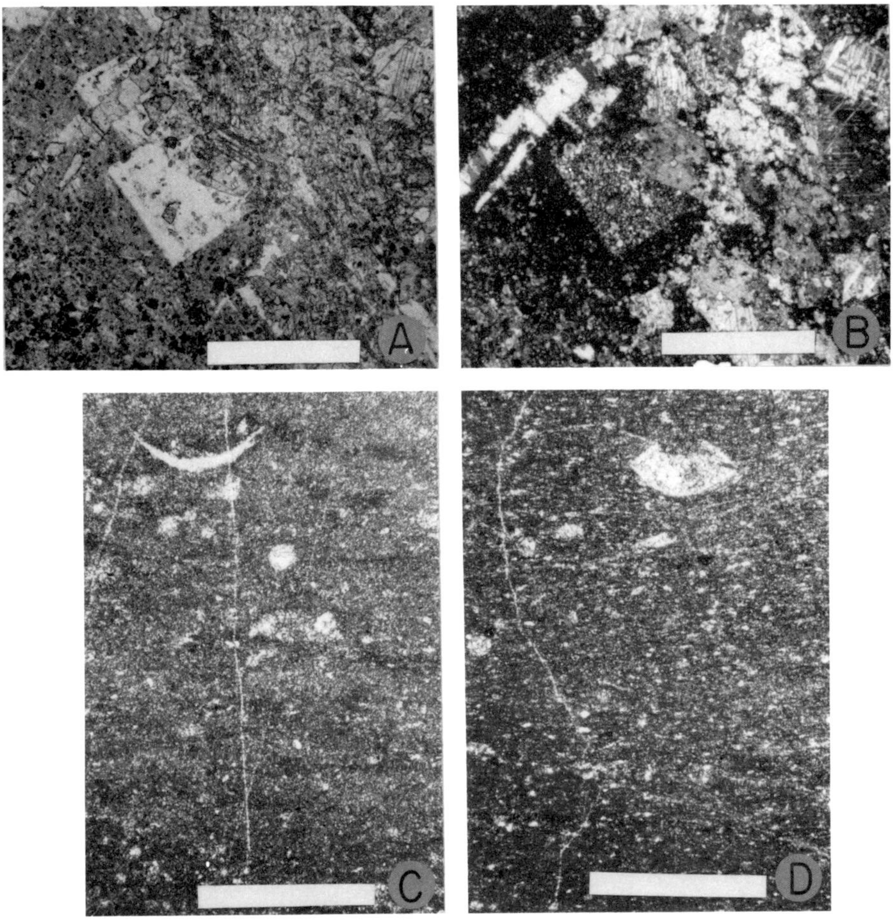

Figure 26. Lenticular dark layers of chert thinly interbedded with platy weathering limestone in an unusually cherty part of the upper member of the Prida Formation, northern East Range.

Figure 27. Lens of massive, coarse sedimentary breccia (left of hammer) in slump-folded, laminated limestone of the upper member of the Prida Formation. East-central part of Humboldt Range.

are generally absent, and fine lamination is preserved. Larger invertebrate fossils are scarce and are almost exclusively pelagic halobiid bivalves and cephalopods.

The thickness of the upper member of the Prida Formation varies inversely with that of the overlying platform carbonate rocks of the Star Peak Group (originally assigned to the "Natchez Pass Formation"), as discussed by Silberling and Wallace (1969, p. 28-29). The upper Prida thins to the east and south away from the northern Humboldt Range. This is because the coordinate Star Peak platform rocks evidently prograded regressively out over the basinal deposits of the upper Prida in a generally northwesterly direction. Tectonic complications preclude accurate stratigraphic measurements in the northern Humboldt Range, but in the northwesternmost of its exposures of the upper Prida where the member has its greatest thickness, it is at least several hundred metres thick. The upper member here has yielded ammonites of probably early Karnian age (Silberling and Wallace, 1969, p. 20) that are correlative with those found near the top of the Augusta Mountain Formation of the platform facies.

The upper member of the Prida Formation gradationally overlies the Fossil Hill Member which also is a basinal deposit. Compared to the Fossil Hill, however, the upper member is more uniform in layering and composition. The small amount of fine-grained terrigenous clastic material is disseminated through the limestone rather than being interstratified in discrete units of calcareous siltsone and shale. Moreover, the average thickness of Triassic ammonite zones within the upper member is about an order of magnitude greater than those within the Fossil Hill in the western region, suggesting a much greater rate of sediment accumulation for the upper member. The relative contribution of terrigenous clastic and pelagic-organic sediments to the upper member and the Fossil Hill Member may have been about the same, but the difference in sedimentation of the upper Prida resulted from a much greater and consistent supply of fine-grained carbonate sediment derived from the nearby carbonate platform. Hence, the transition from the Fossil Hill upward into the upper member of the Prida, which coincidentally approximates the Anisian-Ladinian interstage boundary, may document the initial development of a carbonate-platform depositional complex during Star Peak deposition.

In addition to southward and eastward wedging out of the upper Prida by depositional progradation of the platform margin, the upper parts of the more southern of the upper Prida sections in the northern East Range, and those in the east-central part of the Humboldt Range, have been thoroughly altered to sugary dolomite. This dolomite replaces all of the carbonate rocks to depths as great as 100 m beneath the Panther Canyon Member of the Augusta Mountain Formation (Nichols, 1974). The succession of rock units in these places, in ascending order, is upper Prida, secondary dolomite, and Panther Canyon. Some of the upper Prida may have been eroded away during local uplift in Ladinian time so that the secondary dolomite beneath the Panther Canyon may represent the entire thickness of upper Prida strata. However, the original nature of the dolomitized strata is indeterminate. Even beyond the areal limits of the Panther Canyon Member in the Humboldt Range, scattered rhombs and patches of secondary dolomite are widespread in the upper Prida. Some of the partly dolomitized basinal limestone that was affected by solutions of probable evaporitic origin manifest a complex diagenesis involving replacement first by evaporite minerals and then by silica and dolomite (Figs. 25A, 25B).

Congress Canyon Formation

The name "Congress Canyon Formation" is proposed here for the coarse-grained, allochem-supported limestone interpreted to have been deposited at the platform margin between the depositional sites of the basinal rocks that form the upper member of the Prida Formation and the more inner platform rocks included in the Augusta Mountain Formation. The type section is the virtually unbedded and clifflike exposure of limestone at the head of Congress Canyon in the northern Humboldt Range (Fig. 28), which was mapped as the "lower member of the Natchez Pass Formation" by Wallace and others (1969b). These rocks have undergone

Figure 28. Cliffs formed by typical exposures of the Congress Canyon Formation at the head of Congress Canyon (left side of photograph), Humboldt Range. View north toward summit of Star Peak.

hydrothermal recrystallization, so that the original microscopic character of the skeletal fragments and matrix is mostly obliterated (Fig. 18D). Nevertheless, in hand specimen skeletal fragments of colonial coral and solenporoid(?) algae as much as several centimetres across can be recognized among other coarse shelly and echinodermal grains which make up the massive limestone in the lower part of the Congress Canyon. The contrast is striking between these rocks and the well stratified, fine-grained limestone of the underlying Prida Formation. Higher strata within the limestone of the Congress Canyon cliffs are not so conspicuously coarse grained and may contain interstratified units of recrystallized lime packstone and wackestone like those of the Smelser Pass Member of the Augusta Mountain Formation. However, the bedding and mottling characteristic of the Smelser Pass are not preserved in the Congress Canyon rocks. Therefore, the entire thickness of cliff-forming massive limestone, formerly regarded as the "lower member of the Natchez Pass Formation," should be mapped with the Congress Canyon Formation throughout the northern Humboldt Range.

The Congress Canyon Formation is poorly represented outside the northern part of the Humboldt Range. This is so even though in the western Star Peak region the basinal limestone of the upper member of the Prida Formation underlies platform carbonate rocks of the Augusta Mountain

Formation that prograded regressively out over the depositional basin. In keeping with the facies model for carbonate-platform deposits, strata such as those of the Congress Canyon, that represent the platform margin, should occur in sequence between the upper Prida and the Augusta Mountain. However, in the northern East Range and in some parts of the Humboldt Range pervasive secondary dolomitization beneath the primary dolomite of the Panther Canyon Member of the Augusta Mountain Formation has obliterated features of the transition from basinal to platform carbonate rocks. The resulting geologic map pattern is like that in the northern East Range (Fig. 3B) where featureless, sugary, secondary dolomite separates the upper member of the Prida from the Panther Canyon Member. In this particular area the secondary dolomite is probably too thin in places for it to have replaced all of the strata that were deposited at and near the platform margin. Therefore, these strata could have been eroded away prior to dolomitization and deposition of the Panther Canyon. Similar stratigraphic relations occur also in the central part of the Humboldt Range. In the low hills at the range front between Indian Creek and American Canyon, the unit mapped as the "lower member of the Natchez Pass Formation" by Wallace and others (1969b) is cut be several small faults and comprises massive sugary secondary dolomite and the Panther Canyon and Smelser Pass Members of the Augusta Mountain Formation, as defined here. Only at the south end of these exposures in American Canyon (Fig. 3A) are rocks that may represent the platform margin preserved beneath the massive secondary dolomite and above the upper Prida. The usual dark, evenly stratified limestone of the upper Prida here is overlain by about 10 m of carbonate conglomerate and sedimentary breccia which were derived in part from the upper Prida. These rocks grade upward into about 15 m of current-bedded, coarse shelly lime grainstone and sedimentary breccia which gradually are transformed into the overlying massive sugary dolomite. The coarse-grained deposits overlying the upper Prida are unlike the massive skeletal limestone of the type Congress Canyon but are included in this formation because of their high-energy, platform-margin character. Exposures of this part of the section still farther south in the Humboldt Range, near Buffalo Mountain, are beyond the geographic limits of the Panther Canyon primary dolomite and its attendant secondary dolomitization and therefore show the transition from basinal to platform rocks. Here, however, only about 15 m of light-gray, massive, coarsely skeletal-clastic limestone resembling the Congress Canyon separates the upper member of the Prida from the medium- to thick-bedded, massive or mottled, lime mud-supported rocks of the Smelser Pass Member. The Congress Canyon would thus be too thin to be recognized here as a separate formation at most map scales.

Age-diagnostic fossils are absent in the sparse exposures of the Congress

Canyon Formation. In some places, however, the upper member of the Prida Formation is equivalent in age to most or all of the Augusta Mountain Formation. Therefore, if the Congress Canyon is inferred to have formed at the platform margin between the depositional sites of these two other units, it must be strongly time transgressive. The Congress Canyon Formation would mark the position of the northwesterly prograding margin of the Star Peak carbonate platform and range in age from early Ladinian through the late Ladinian and probably into early Karnian, as shown on Figures 2 and 31. Except in the northern Humboldt Range, where the formation has an appreciable thickness, Congress Canyon deposition at any one time would have been confined to a relatively narrow belt along the edge of the carbonate platform; the amount of time represented by the Congress Canyon section at any one place would be brief owing to progradation of the platform.

Augusta Mountain and Cane Spring Formations

In the western Star Peak outcrop region, the Cane Spring Formation, the rocks named by us as Smelser Pass Member of the Augusta Mountain Formation, and in places the Panther Canyon Member of the Augusta Mountain Formation have been included in the "Natchez Pass Formation" of Ferguson and others (1951a), a formational name that we are abandoning. Not only does the concept of the "Natchez Pass" needlessly overlap natural stratigraphic units that were recognized and in part previously named elsewhere among strata now included within the Star Peak Group, but the base of its typical exposures in the northern East Range corresponds to the lower limit of the secondary dolomite beneath the Panther Canyon (Fig. 3B). Thus, the base of the "Natchez Pass Formation," as originally defined, is not a primary depositional boundary. The original application of the name "Natchez Pass Formation" in the vicinity of its type area in the northern East Range is delineated by the double-ended arrow in Figure 3B.

The type locality of the "Natchez Pass Formation" was designated by Ferguson and others (1951a) as being at Natchez Pass, about 5 km north of the area represented in Figure 3B. The formation here contains the same rock units as those shown in Figure 3B, but its base is not exposed. Farther north in the northern East Range, at the type locality originally specified for the Prida Formation, dynamothermal alteration has obscured the Prida-"Natchez Pass" contact and the nature of the carbonate rocks that would represent the lower part of the "Natchez Pass Formation." Thus, although it is not designated as the type locality of either the Prida or "Natchez Pass," the general area shown in Figure 3B is the only part of the pre-Tertiary outcrop area in the northern East Range where the stratigraphy of these rocks is adequately known.

Where present in the western Star Peak region, the Panther Canyon Member is the basal part of the Augusta Mountain Formation, but it differs in character from the Panther Canyon farther east. Dense, blocky weathering, light-colored, algal-laminate dolomite, like that which typically composes most of the member's carbonate rocks, is interstratified in units as much as several metres thick with an equal or greater amount of sugary secondary dolomite. Some of the sugary dolomite is homogeneous, resembling that forming the body of secondary dolomite that everywhere lies beneath the Panther Canyon. In other occurrences, the secondary dolomite is darker gray and retains traces of original sedimentary structures or allochems. The dolomite evidently represents original intertidal or lagoonal platform limestone. The stratigraphic limits of the Panther Canyon in the sections that are interpreted as intertongued supratidal and more open platform deposits (for example, see Figs. 3A, 3C) are taken to include the lowest and highest beds of dense, light-colored dolomite. Clastic rocks are generally absent in these sections of the Panther Canyon, but locally as much as a few metres of fine-grained chert pebble conglomerate mark the top of the member. The transition upward from units of dark-gray secondary dolomite to dense, light-colored, commonly algal-laminate "primary" dolomite is gradational, representing the regressive progradation of supratidal deposits out over the carbonate platform. In contrast, the contact between "primary" dolomite and overlying secondary dolomite is sharp and marked commonly by solution or erosional effects such as rip-up breccias of "primary" dolomite fragments in a matrix of dark secondary dolomite. The top of each supratidal "primary" dolomite unit therefore commonly shows evidence of having been subaerially exposed before further relative subsidence and burial beneath originally calcareous sediments of more open-platform character.

The Smelser Pass Member of the Augusta Mountain Formation in the western Star Peak outcrop region is like the typical exposures farther east in being a monotonous succession of mainly mud-supported limestone. The characteristic regular but nodular bedding and bioturbation(?) mottling of the Smelser Pass is not pronounced, however, perhaps because of the generally greater degree of metamorphic recrystallization in the western part of the region. In its northernmost exposures in the East Range, and at places in the Humboldt Range, the Smelser Pass has been altered to coarse-grained marble, some of which is distinctly foliated. Where original fabrics are preserved, allochems in limestone of the Smelser Pass in the western Star Peak region are principally sparry calcite mosaics pseudomorphous after shell fragments, echinodermal grains, or ooids. The oolitic grainstone (Fig. 18C) immediately above the rocks assigned to the Panther Canyon is in the central Humboldt Range included in the Smelser Pass (Fig. 3A). This grainstone evidently represents a tongue of the oolite shoal deposits from which were derived the abundant ooids that were swept into the muddy, more inner platform environments farther east. As such, this grainstone would be essentially a thin tongue of the platform-margin Congress Canyon Formation. This suggests how complex intertonguing may be among correlative parts of the Congress Canyon Formation and the Smelser Pass and Panther Canyon members near the platform margin. Unfortunately, these relations are nowhere well displayed owing to secondary dolomitization, metamorphism, and lack of exposure.

Mafic volcanic rocks generally are restricted to only the uppermost part of the Smelser Pass Member. At the type locality of the member at China Mountain, they are preserved only locally to thicknesses up to several metres. Apparently they were eroded away in most places beneath the unconformity at the base of the overlying Cane Spring Formation. In other areas, however, such as the southern Tobin Range and the Stillwater and Humboldt Ranges, the volcanic rocks form a persistent unit as much as several tens of metres thick. At the south end of the Humboldt Range, in its southeasternmost outcrops, the character of the Smelser Pass is drastically changed by the appearance of abundant mafic volcanic (andesitic?) flows, breccia, and tuffs which intertongue throughout the section. The map pattern of these volcanic rocks is shown by Wallace and others (1969a), and they are described more fully by Silberling and Wallace (1969, p. 22, 29–30). An increase in the amount of mafic volcanic rocks in the "Natchez Pass Formation," probably in the part now assigned to the Smelser Pass, has also been observed by R. C. Speed (1969, oral commun.) in the southernmost Star Peak outcrops in the Stillwater Range.

The Cane Spring Formation in the western and northwestern parts of the Star Peak region is generally like that farther east and shows some of the same patterns of lateral and vertical variation. Locally, however, the carbonate rocks are metamorphosed to coarse-grained calcitic and dolomitic marble, especially in the northern Humboldt and East Ranges.

Chert conglomerate and finer grained siliceous clastic rocks at the base of the Cane Spring Formation are restricted to the more northern exposures in the western part of the region. These clastic rocks contain the only age-diagnostic fossils found thus far in the formation. A bivalve and cephalopod fauna of the Dilleri Zone of early late Karnian age described by Silberling (1961) occurs near the base of a clastic section that measures more than 100 m thick at the northeastern tip of the Humboldt Range. These clastic strata were only tentatively correlated by Silberling and Wallace (1969, p. 30) with the "upper member of the Natchez Pass Formation." Additional knowledge of the sections found in ranges farther east, makes assignment of this unit to the Cane Spring Formation inescapable. In the northern East Range the thickness and composition of the Cane Spring basal clastic strata are highly variable; these rocks may have been deposited on an irregular erosional surface. Their maximum thickness again exceeds 100 m. From argillite in the upper part of the Cane Spring basal clastic rocks here, D. H. Whitebread of the U.S. Geological Survey collected an impression of a clydonitacid ammonite that most closely resembles the genus *Traskites* (sensu lato) which also is indicative of the Dilleri Zone.

Farther south in the northern Stillwater and southern Humboldt Ranges, basal clastic rocks are absent. Brownish-weathering, impure, shelly limestone forms the basal part of the Cane Spring Formation, as it does in the Tobin Range and at places in the Augusta Mountains. This evenly bedded, slightly impure limestone in the southern Humboldt Range was separated by Wallace and others (1969a) as a map unit distinct from the gradationally overlying, more massive and thickly bedded gray limestone that forms the upper part of the formation. These two informal units, each of which is about 150 m thick, together constituted the "upper member of the Natchez Pass Formation" there, and as such they correspond to the Cane Spring Formation. Petrographically, most of the limestone beds in these two units are skeletal molluscan-echinodermal wackestone. The lower unit is distinguished by a small amount of clay and silt impurity, and as a consequence it has undergone less metamorphic recrystallization.

In the southern Humboldt Range, original fabrics are fairly well preserved in some limestones of the Cane Spring. As in the limestone farther east, they characteristically show differential diagenetic solution of aragonitic, as compared with calcitic, skeletal allochems. This fabric suggests the influence of meteoric waters during diagenesis. One thin section in addition shows laminated calcite crusts coating original postlithification surfaces; these closely resemble the crusts developed on exposed surfaces of Pleistocene limestone in the Florida-Bahamas region (Multer and Hoffmeister, 1968).

Other indications of shallow water depths and possible subaerial exposure occur in the upper Cane Spring Formation in the northern Stillwater and East Ranges. In these places the upper several tens of metres of the Cane Spring section include algal-laminate, fine-grained dolomite, and at one place in the Stillwater Range, patches of coarsely crystalline calcite in secondary dolomite may be pseudomorphous after sulfate-mineral replacement nodules (Fig. 29). The geographic distribution of these supratidal and evaporitic rocks coincides with the western boundary of the older intra-Star Peak uplift, of Ladinian age, and suggests renewed emergence along this same trend. In the northernmost exposures in the East Range, the metamorphosed Cane Spring carbonate rocks are much reduced in thickness to only about 100 m. Some of this thinning may be tectonic, as these beds are involved here in tight large-scale folds, but thinning could also result from local emergence and erosion prior to deposition of the Grass Valley Formation (Figs. 4, 5).

Figure 29. Patches and stringers of white crystalline dolomite which possibly replace evaporitic sulfate minerals in gray sugary dolomite associated with laminated ''primary'' dolomite. Cane Spring Formation, northern Stillwater Range.

CORRELATIVE STRATA BEYOND THE STAR PEAK OUTCROP AREA

The Star Peak exposures in and around the Sonoma Range one-degree quadrangle are a coherently interrelated natural entity, albeit a stratigraphically complex one. Beyond this area, in nearby parts of western Nevada, exposures of correlative rocks are scarce. Only in the outcrop area of the ''Luning sequence'' (Silberling and Roberts, 1962), 100 to 150 km south of the Star Peak area, are Triassic carbonate strata widely exposed, but most of these rocks are younger than the Star Peak Group. In the Luning sequence, platform carbonate rocks correlative with those of the Star Peak Group occur only in the lowest part of the Luning Formation and in the underlying Grantsville Formation. The Grantsville, which is recognized only in the Shoshone Mountains and forms the oldest part of the Luning sequence, correlates with the upper part of the Augusta Mountain Formation of the Star Peak Group. The Grantsville disconformably underlies the Luning Formation, the basal part of which correlates with the Cane Spring Formation. Thus, the mid-Karnian episode of uplift and erosion recorded in the upper part of the Star Peak Group evidently affected the carbonate platform farther south as well. The andesitic volcanic and associated sedimentary rocks beneath the Luning sequence in the Shoshone Mountains were erroneously assigned to the Pablo Formation, of supposed Permian age, by Silberling (1959). They are still poorly dated but could be a southern analog of the Koipato Group and be as young as the lower part of the Star Peak Group (Silberling, 1973, p. 355). However, these rocks are of limited areal extent, and most of the carbonate platform complex of the overlying Luning sequence is younger than any part of the Star Peak Group.

Isolated exposures of strata outside the Star Peak region that are at least in part genetically related to the Star Peak Group occur only in the Shoshone, Toiyabe, and New Pass Ranges.

Shoshone Range

Strata of the Star Peak Group in the northwestern Shoshone Range, just east of the main Star Peak outcrop area, were first described by Gilluly and Gates (1965) who tentatively assigned them to the China Mountain Formation. They also recognized that these rocks lie unconformably on strata of both the Havallah Formation and Antler Peak Limestone, a relationship requiring that they overlap the Golconda thrust which regionally juxtaposed the Havallah and related rocks with the Antler Peak and its depositionally related rocks. Nichols (1971) confirmed this relationship and showed these overlapping strata to be the distinctive dolomite and clastic rocks of the Panther Canyon Member of the Augusta Mountain Formation rather than part of the China Mountain Formation. In the Panther Canyon section here, as in the vicinity of China Mountain, a basal unit of yellow-weathering dolomite is overlain by clastic rocks, including maroon conglomerate; the total thickness is about 50 m. No other rocks of the Star Peak or Koipato Groups are present. These strata, having a total outcrop area in the Shoshone Range of less than 1 km^2, are of significance because they demonstrate that movement on the Golconda thrust preceded deposition of the Star Peak Group, as well as the nearly concordant rocks of the still older Koipato Group. They indicate also that older parts of the Star Peak Group, along with the Koipato Group, either were not laid down so far east or else were removed by erosion prior to deposition of the Panther Canyon.

Toiyabe Range

In the northernmost Toiyabe Range, near the 40th Parallel, more than 10 km^2 of Triassic outcrop were first discovered by J. H. Stewart in 1969 and are shown on the geologic map of Lander County, Nevada (Stewart and McKee, 1977). These are the most easterly exposures of marine Middle or Upper Triassic rocks known in Nevada. Most of these rocks in the Toiyabe Range are monotonously similar. Exposures are generally poor and are cut by numerous faults, so that the maximum exposed thickness of these rocks is unknown. However, at least a few hundred metres of distinctive, generally medium- or thick-bedded, olive-gray weathering, mainly lime-mud supported limestone is represented. This limestone grades downward through impure calcareous rocks into basal conglomerate, as much as tens of metres thick, which rests unconformably on rocks of the Golconda allochthon. The lower part of the limestone in places is notably shelly and contains a diverse fossil fauna of brachiopods, gastropods, bivalves, corals, and sponges. A trachyceratid ammonite—either *Trachyceras* or *Daxatina*—from these beds documents a latest Ladinian or earliest Karnian age, equivalent to part of the Smelser Pass Member of the Augusta Mountain Formation in the Star Peak Group. However, these isolated rocks in the Toiyabe Range are too different lithologically from the Smelser Pass to be considered part of it. Neither are they like the age-equivalent carbonate rocks in the New Pass Range. In general, the Triassic rocks in the Toiyabe Range have more in common with the correlative Grantsville Formation which crops out about 120 km southwest of them in the Shoshone Mountains.

In keeping with the open-marine character of the invertebrate fauna in the limestone from the Toiyabe Range, some beds low in the section are coarse-grained crinoid or algal, intraclast grainstone (Fig. 30), which indicates an energetic carbonate depositional environment. Why this is so for these strata that are much farther east, and generally more inshore, than the correlative platform rocks farther west is inexplicable. It is, however, not so contradictory relative to the correlative basinal limestone in the New Pass Range, not far south of the Star Peak outcrop area. A complicated paleogeography is indicated, whereby basinal and outer platform carbonate environments existed southeast and south as well as west of the Star Peak depositional site; or it may be that major geographic rearrangement has resulted from hitherto unrecognized post-Triassic tectonic activity.

New Pass Range

Calcareous Triassic strata that resemble and are correlative with the Star Peak Group crop out extensively in the New Pass Range only 20 to 30 km south of the characteristic Star Peak exposures in the Augusta Mountains. These rocks could reasonably be included in the group, but of the various formational names applied to the Star Peak, only the name "Favret Formation" can properly be used in the New Pass Range. Intervening exposures that would relate the New Pass Range section to the Star Peak stratigraphy are lacking.

The early attempt by Silberling (1956, p. 1151) to apply in the New Pass Range the formational terminology developed by Muller and others

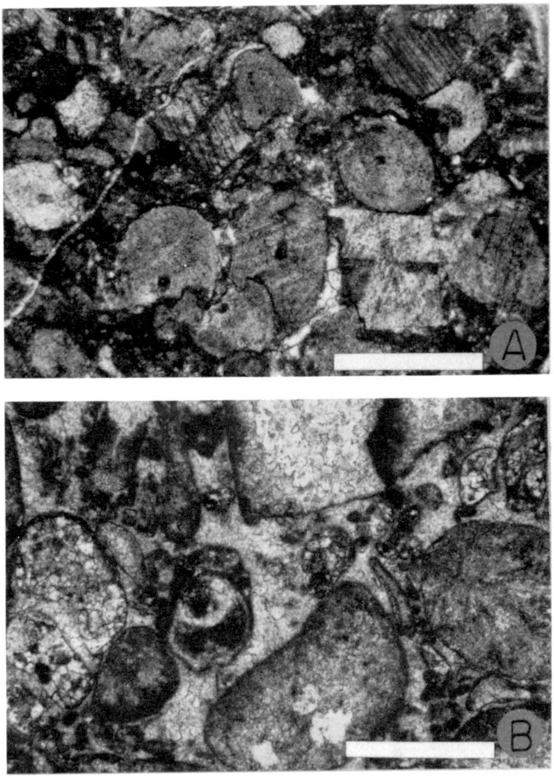

Figure 30. Photomicrographs of Upper Triassic rocks near Hall Creek, northern Toiyabe Range. A, stylolitized crinoid biomicrite. Pelletlike grains may be completely micritized shell fragments. Bar = 0.50 mm. B, algal(?)-molluscan biosparite. Note well-developed micritic envelopes. Clear patches at lower center are length-slow chalcedony replacements in algal(?) grains. Bar = 0.50 mm.

(1951) and Ferguson and others (1952) farther north in the Sonoma Range quadrangle is largely erroneous. Mosher (1968, p. 899) and Willden and Speed (1974, p. 6) were misled into following the same usage. The principal error in this terminology resulted from not realizing that the middle part of the type Augusta Mountain Formation in the Augusta Mountains corresponds to the Panther Canyon member and that these strata are absent in the New Pass Range. Instead, none of the rock units above the Favret Formation in the New Pass Range have close lithologic counterparts farther north, nor is the name "Dixie Valley Formation," as applied in the New Pass Range by Silberling (1956), appropriate for the rocks underlying the Favret Formation.

If we are correct in regarding the sub-Favret clastic and tuffaceous rocks in the New Pass Range, including the "conglomerate unit" of Willden and Speed (1974, p. 6), as counterparts of the upper Koipato Group, then the Favret Formation represents the oldest part of the Star Peak Group in this area. At least a few metres of massive, shelly limestone at the base of the Favret correspond to the lower member of the formation farther north. The overlying Fossil Hill Member, about 200 m thick, contains near its base mid-middle Anisian ammonites representative of the upper part of the *Acrochordiceras hyatti* beds in the zonation of Tozer (1971). Higher parts of the Fossil Hill yield ammonite and pelagic bivalve fossil faunas representing the rest of the middle Anisian and, as elsewhere, the entire upper Anisian. About 500 m of mostly massively bedded, boldly outcropping gray limestone overlies the Favret. Approximately the lower half of this limestone contains much interstratified calcareous siltstone, limestone with carbonate-supported quartz grains and chert pebbles, and beds of siliceous sandstone and pebble conglomerate. The cleaner limestone beds are noticeably rich in organic-skeletal debris. Because of its clastic component, this part of the section could possibly be an undolomitized equivalent of the Home Station Member of the Augusta Mountain Formation. Sand-size and coarser terrigenous detritus is absent higher in the limestone sequence, where coarse calcareous skeletal grains form the major share of the rock. Organic buildups of colonial coral and large molluscan shells occur locally and are associated with crinoidal and algal grain-supported limestone and some oolitic beds. Layers packed with oncolites and large fragments of solenoporoid(?) algae occur in less cleanly washed limestone. These open-marine bioclastic rocks have no lithic counterpart in the Star Peak outcrop area, but they are probably correlative in age with the supratidal dolomite and clastic rocks of the Panther Canyon Member of the Augusta Mountain Formation. An abrupt change takes place above them to thin-bedded, partly laminated and flaggy weathering dark-gray limestone and fine-grained calcareous clastic rocks which total about 200 m in thickness. These rocks of open-marine, basinal character contain an abundant ammonite and halobiid-bivalve fossil fuana, and their occurrence here constitutes the type example of the Desatoyense Zone (Silberling and Tozer, 1968, p. 35) of earliest Karnian age. Thus, these basinal rocks correlate, at least in part, with the lithologically different platform limestone of the Smelser Pass Member of the Augusta Mountain Formation. The highest exposed 200 m of the section in the New Pass Range is again composed of massive, cliff-forming, bedded limestone and dolomite which could be equivalent to the Cane Spring Formation at the top of the Star Peak Group. Further description of these rocks in the New Pass Range is provided by MacMillan (1972).

Northwestern Nevada

Except in the far northwestern corner of the state, rocks as old as the Star Peak Group are nowhere known in Nevada to the southwest, west, and north of the Star Peak outcrop area. In the Pine Forest Range, about 100 km northwest of the nearest Star Peak exposures, fossils of probable Middle or Late Triassic age are reported by Willden (1964, p.

49) from a thick metamorphosed sequence of limestone, conglomerate, sandstone, and metapelitic rocks. These rocks overlie the dominantly andesitic Happy Creek Group of probable Permian age. The nearby Quinn River Formation (Willden, 1964, p. 41-42), which consists of about 200 m of "silty shale" with interbeds of impure limestone and chert, in a single small exposure, has yielded more age-diagnostic late Middle Triassic fossils. The paleogeographic setting of these Triassic rocks is unresolved, but they are associated with Permian strata that have affinities with rocks in the eastern Klamath Mountains (Silberling, 1973, p. 358); they bear no apparent relation to the Star Peak Group.

Depositional History

The depositional history of the Star Peak Group is complicated by coeval tectonic activity within the site of sedimentation. As the lithologic nature of platform carbonate rocks, such as those of the Star Peak, is especially sensitive to relative changes in sea level, episodes of tectonism during Star Peak deposition resulted in striking facies changes different from those that would develop by continuous deposition in a generally subsiding marine basin. Although the geologic events recorded in the Star Peak rocks characterize only a relatively small area, they provide information on the nature of the continental margin during a critical time of apparent change in tectonic style at this general latitude in western North America. Prior to deposition of the Star Peak Group, the tectonic record of much of Paleozoic and earliest Triassic time has been characterized as one of developing and collapsing marginal (or "inner arc") ocean basins (Burchfiel and Davis, 1972; Churkin, 1974; Silberling, 1973), whereas during or after Star Peak time, the Sierra Nevada magmatic arc became established at the edge of an "Andean-type" continental margin (Schweickert and Cowan, 1975). The actual record of tectonic events during the middle part of Triassic time is meager even though some first-order structural events have been postulated for that time, such as truncation of the continental margin (Burchfiel and Davis, 1972) and large-scale strike-slip rifting (Jones and others, 1972; Silver and Anderson, 1974).

The depositional and tectonic history of the oldest parts of the Star Peak Group have been obscured by subsequent intra-Star Peak uplift, erosion, and secondary dolomitization. During late Spathian time, open-marine calcareous strata may have been deposited upon a fairly flat surface throughout much of the western and southern part of the Star Peak outcrop area, as if the entire region were abruptly inundated by the sea. However, only remnants of such strata are preserved in the lower member of the Prida Formation and in the Tobin Formation. Stratigraphic evidence such as that in the Humboldt Range (Fig. 22) indicates that local relative uplift and erosion took place during early Anisian time, although the nature of the uplift—whether by faulting or folding—is not clear. Farther to the southeast, the coarse clastics in the Dixie Valley Formation were evidently shed southward from a highland within the Star Peak area, also documenting relative uplift during the early Anisian. Too little of the record is preserved, however, to determine the shape and size of the area or areas that underwent relative uplift at that time.

Deposition of the basinal marine calcareous deposits of the Fossil Hill Member of either the Prida or Favret Formations began during middle Anisian time in subsided parts of the Star Peak area, and by late Anisian time Fossil Hill rocks blanketed the entire region. These deposits overlapped the previously uplifted areas as if the entire region had again undergone relative subsidence. The known and inferred distribution of the Fossil Hill is shown in Figure 31; platform or nearshore carbonate equivalents of these basinal rocks are unknown.

Although the older parts of the Star Peak Group are mainly calcareous rocks, the first record of carbonate-platform construction is found in rocks of early Ladinian age. Figures 31B through 31F diagrammatically show

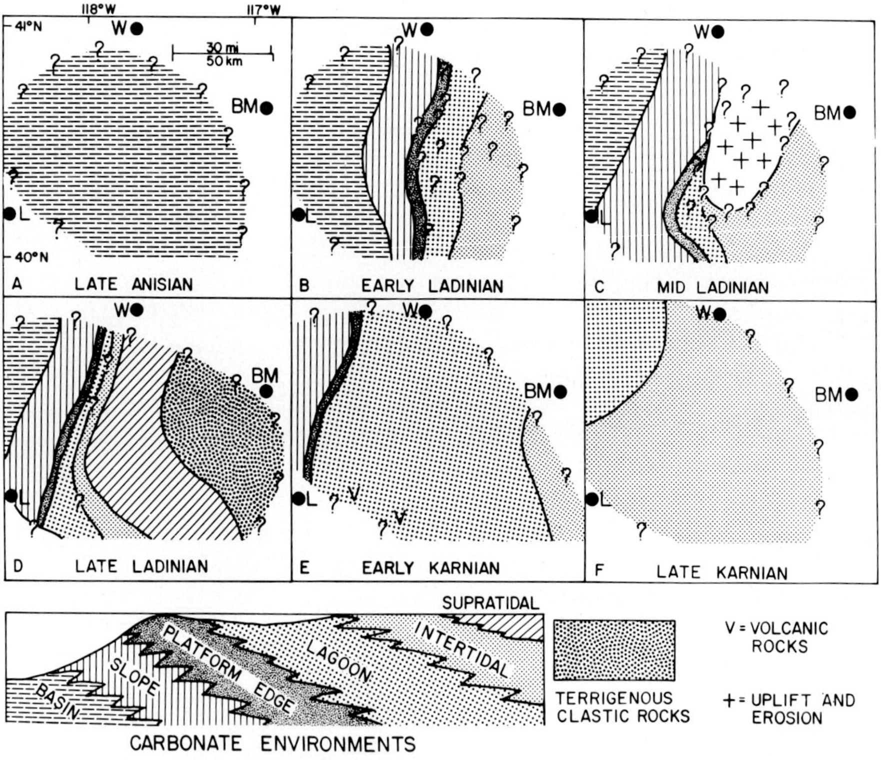

Figure 31. Paleogeographic maps of the Star Peak Group at different times during its deposition. See Figure 5 for series assignment of Triassic stages. Area of maps is approximatley that in Figure 1; L, Lovelock; W, Winnemucca; BM, Battle Mountain. The map patterns correspond to those on the conventional model of near-shore carbonate environments shown beneath the maps.

the further development of this platform to the end of Star Peak deposition in late Karnian time.

In the western Star Peak region, open-marine Anisian strata of the Fossil Hill Member grade upward into dark, laminated, cherty limestone of the upper member of the Prida Formation. This progression represents a continuation of basinal deposition during Ladinian time concomitant with uplift and erosion farther east in the central part of the present Star Peak outcrop area (Figs. 31B, 31C). Existence of a slope environment between the basin and platform is shown by sedimentary breccia and soft-sediment deformation within the upper Prida section. South and east of the uplifted area, subsidence and peritidal deposition of the carbonate and clastic rocks of the Home Station Member of the Augusta Mountain Formation took place. A topographic break must have separated the upper Prida basinal deposits from those of the Home Station, but its nature is obscured by erosion of the uplifted area and associated secondary dolomitization, as indicated by the liberal use of question marks on the reconstructed paleogeogrpahy shown in Figure 31C. Little actual record is preserved of the lower or middle Ladinian platform margin whose position evidently was tectonically controlled because it coincides with the western edge of the uplifted area. Although erosion during Star Peak time exposed pre-Star Peak rocks on the uplifted platform, coarse-grained terrigenous clastic material was not shed westward into the basinal upper Prida deposits. Siliceous sand, grit, and pebbles are sporadically scattered through the correlative Home Station Member carbonate rocks and may have been derived from the platform which therefore may have been an eastwardly tilted block. This geometry is also suggested by progressively deeper erosion of Fossil Hill strata beneath the Ladinian unconformity in a westerly and northwesterly direction at China Mountain and in the southern Tobin Range, respectively. However, the nature of the uplift, whether by faulting or folding, is unknown.

By late Ladinian time the uplifted area was beveled by erosion, and the entire eastern and central Star Peak regions became a vast supratidal flat upon which the Panther Canyon Member of the Augusta Mountain Formation accumulated under evaporitic conditions. First the algal-laminate "primary" dolomite of the Panther Canyon formed, and then, more locally, the Panther Canyon clastic sediments spread over part of the carbonate platform from the northeast. Regionally extensive secondary dolomitization of older Star Peak limestone took place just prior to or during Panther Canyon deposition (Nichols, 1972, 1974). Supratidal conditions alternated with more open-platform intertidal or lagoonal environments along the west and southwest side of the platform. This was manifiest by intertongues in the Panther Canyon of coarse-grained secondary dolomite in the northern East Range and in the Stillwater Range. The position of the platform margin is fairly well located geographically for this time (Fig. 31D), and actual platform margin deposits crop out in the southern Humboldt Range. In the northern part of the Star Peak area, the narrowness of the belt in which the more outer-platform deposits of this age occur may result in part from telescoping of facies by low-angle post-Triassic faults such as those of the Clear Creek system of thrust faults (Silberling, 1975) which have displacements of as much as a few tens of kilometres.

Further subsidence during the early Karnian allowed the platform limestone of the Smelser Pass Member of the Augusta Mountain Formation to thickly blanket the region. At the same time the platform margin, represented by the Congress Canyon Formation, built regressively westward out over the persistent basinal deposits of the upper member of the Prida Formation (Fig. 31E). A paleogeographic complication during this time is the prevalence of mafic volcanic rocks in the Smelser Pass Member in the southern Humboldt and Stillwater Ranges. In that part of the region the Star Peak platform may have been supported by a local volcanic center (Silberling and Wallace, 1969, p. 30). Unfortunately, no exposures of Star Peak rocks exist farther west or south, so the paleogeographic setting of these volcanic rocks cannot be fully known. The margin of the platform presumably swung around this volcanic center to the west and south, because basinal deposits correlative with the Smelser Pass occur in the New Pass Range, south of the principal Star Peak outcrop region.

During the middle part of Karnian time, a break in deposition, accompanied by erosion, channeling, and development of solution caves (Nichols, 1972) within the Smelser Pass, took place at least in the northeastern part of the Star Peak outcrop area. This was followed be deposition of coarse conglomerate and other clastic rocks that form the lower part of the Cane Spring Formation. To the west and south these clastic rocks tongue out, but the base of the Cane Spring could nevertheless rest disconformably on the Smelser Pass. Following this brief but widespread tectonic interruption of the carbonate depositional pattern, upper Karnian platform carbonate strata of the Cane Spring Formation blanketed not only the older platform but also extended westward beyong the limits of present-day Star Peak exposures (Fig. 31F). This westward shift of the platform margin in late Karnian time may have extended at least as far as the northwestern Sierra–eastern Klamath belt, where the only significant amount of carbonate rock in the primarily volcanic and clastic Triassic section is the Hosselkus Limestone of late Karnian age. If this shift occurred, the carbonate platform during Cane Spring deposition would have extended at least 100 km beyond the present northwesternmost Star Peak exposures. This distance is based on restoration of the northwestern Sierra-Klamath belt to its possible pre-Tertiary position with respect to northwestern Nevada, following Hamilton (1969, Fig. 4).

Paleotectonic Interpretation

The mid-Karnian interruption in Star Peak deposition was the result of a more widespread tectonic disturbance that affected areas far beyond the Star Peak region. Pronounced uplift and erosion to the north and northeast of the Star Peak region during this time is indicated by the increase, in these directions, in thickness and clast size of the basal conglomerate unit of the Cane Spring Formation. Although the stratigraphy of the Star Peak Group does not further reflect the nature of this tectonic event, it does record an episode of significant regional relative uplift.

Earlier tectonic events during Middle Triassic time were localized within the Star Peak outcrop area and produced stratigraphic patterns from which additional tectonic inferences can be drawn. Relative uplift during the Ladinian in the central part of the region is especially well documented by the Star Peak stratigraphic record. However, the boundaries of the uplifted area are not adequately displayed, so its is uncertain whether the mechanism was faulting or folding. During late Anisian time, prior to local uplift, relatively deep-water, open-marine strata of the Fossil Hill rock unit were laid down across the entire region. Then, following relative uplift, the locally emergent area was eroded while deposition continued to the east, south, and west of it. The basinal upper member of the Prida Formation was deposited to the west and southwest, and the peritidal Home Station Member of the Augusta Mountain Formation was deposited to the east and southeast. In late Ladinian time, the widespread evaporitic dolomite of the Panther Canyon Member accumulated across the uplifted and beveled area.

The earlier Middle Triassic episode of local uplift that took place during early Anisian time, although not so well defined, seems to have had the same general pattern as the subsequent Ladinian uplift and to have been on the same regional scale. Initially, during latest early Triassic time, ammonite-bearing limestone may have been laid down over a wide area, part or parts of which were then uplifted and eroded, only to be blanketed again by open-marine upper Anisian deposits of the Fossil Hill.

This pattern of local relative uplift accompanied elsewhere within the Star Peak region by uninterrupted deposition can be interpreted two ways. The uplifted blocks may have undergone successive episodes of uplift and subsidence, and erosion of them would have taken place when they emerged above sea level, which remained constant. Alternatively, blocks of various sizes may have undergone episodes of differential subsidence during a time of continually lowering sea level so that down-dropping blocks that lagged behind would have had their tops eroded when falling sea level caught up with them. Because uplift and erosion seem to have occurred twice during Middle Triassic time, we prefer the second alternative. The amount of drop in sea level required by this explanation would seem to be too great to be eustatic. Therefore, we suggest that during Middle Triassic time the entire region was broadly upwarped resulting in crustal extension accompanied by the episodic tensional subsidence of local areas of varying size and shape. This paleotectonic interpretation, at least for earlier Star Peak deposition, accords well with that envisaged for the still older Koipato Group which in the southern Tobin Range was found by Burke (1970a, 1973) to have been deposited concurrently with block faulting.

Putting these inferences in larger perspective, the Sonoma orogeny, which resulted in thrust emplacement of the upper Paleozoic oceanic rocks of the Golconda allochthon during latest Permian or Early Triassic time, was followed during the Triassic by regional crustal extension and local differential subsidence during deposition of the dominantly volcanic Koipato Group and carbonate rocks of the Star Peak Group.[1] The abrupt change during mid-Late Triassic time from Star Peak carbonate platform sedimentation to that of the rapidly deposited, terrigenous clastic, pelitic and sandy strata of the overlying Auld Lang Syne Group may relate to another fundamental change in tectonic regime, such as the beginning of the Sierra Nevada magmatic arc.

[1] Speed (1977) has recently suggested that the western and southern limits of lower Mesozoic deposition in northwestern Nevada coincide approximately with those of a vast volcanic arc assemblage which everywhere underlies the lower Mesozoic rocks but rarely crops out. This arc-related assemblage is envisaged as having moved relatively westward toward the continent during the Sonoma orogeny, squeezing out the Golconda allochthon before it. Thermal contraction of this arc is then called upon to produce subsidence of the lower Mesozoic basin. The Star Peak outcrop area constitutes only a small, peripheral part of this basin and may not reflect the principal deep-seated tectonic controls on deposition. Nevertheless, patterns of sedimentation within the Star Peak Group, as described above, require at least local uplift. They cannot be explained by subsidence alone.

References Cited

Burchfiel, B. C., and Davis, G. A., 1972, Structural framework and evolution of the southern part of the Cordilleran orogen, western United States: Am. Jour. Sci., v. 272, p. 97-118.

Burke, D. B., 1970a, Permo-Triassic non-marine deposition and coeval block faulting, west-central Nevada: Geol. Soc. America Abs. with Programs, v. 2, p. 77-78.

——1970b, Reinterpretation of the "Tobin thrust," west-central Nevada: Geol. Soc. America Abs. with Programs, v. 2, p. 78.

——1973, Reinterpretation of the "Tobin thrust": pre-Tertiary geology of the southern Tobin Range, Pershing County, Nevada [Ph. D. Dissert.]: Stanford, Calif., Stanford Univ., 82 p.

Cameron, E. N., 1939, Geology and mineralization of the northeastern Humboldt Range, Nevada: Geol. Soc. America Bull., v. 50, p. 563-634.

Churkin, Michael, Jr., 1974, Paleozoic marginal ocean basin-volcanic arc systems in the Cordilleran foldbelt, *in* Dott, R. H., Jr., and Shaver, R. H., eds., Modern and ancient geosynclinal sedimentation: Soc. Econ. Paleontologists and Mineralogists Spec. Pub. 19, p. 174-192.

Dunham, R. J., 1962, Classification of carbonate rocks according to depositional texture, *in* Ham, W. E., ed., Classification of carbonate rocks: Am. Assoc. Petroleum Geologists Mem. 1, p. 108-121.

Ferguson, H. G., Muller, S. W., and Roberts, R. J., 1951a, Geology of the Winnemucca quadrangle, Nevada: U.S. Geol. Survey Geol. Quad. Map GQ-11.

——1951b, Geology of the Mount Moses quadrangle, Nevada: U.S. Geol. Survey Geol. Quad. Map GQ-12.

Ferguson, H. G., Roberts, R. J., and Muller, S. W., 1952, Geology of the Golconda quadrangle, Nevada: U.S. Geol. Survey Geol. Quad. Map GQ-15.

Folk, R. L., 1959, Practical petrographic classification of limestones: Am. Assoc. Petroleum Geologists Bull., v. 43, p. 1-38.

——1962, Spectral subdivision of limestone types, *in* Ham, W. E., ed., Classification of carbonate rocks: Am. Assoc. Petroleum Geologists Mem. 1, p. 62-84.

Folk, R. L., and Pittman, J. S., 1971, Length-slow chalcedony: A new testament for vanished evaporites: Jour. Sed. Petrology, v. 41, p. 1045-1058.

Gilluly, James, and Gates, Olcott, 1965, Tectonic and igneous geology of the northern Shoshone Range, Nevada: U.S. Geol. Survey Prof. Paper 465, 153 p.

Goddard, E. N., chm., and others, 1948, Rock-color chart: Washington, Natl. Research Council (repub. by Geol. Soc. America, 1963), 6 p.

Hamilton, Warren, 1969, Mesozoic California and the underflow of Pacific mantle: Geol. Soc. America Bull., v. 80, p. 2409-2430.

Horowitz, A. S., and Potter, P. E., 1971, Introductory petrography of fossils: New York, Springer-Verlag, 302 p.

Ingram, R. L., 1954, Terminology for the thickness of stratification and parting units in sedimentary rocks: Geol. Soc. America Bull., v. 65, p. 937-938.

Johnston, F. N., 1941, Trias at New Pass, Nevada (new lower Karnic ammonoids): Jour. Paleontology, v. 15, p. 447-491.

Jones, D. L., Irwin, W. P., and Ovenshine, A. T., 1972, Southeastern Alaska—a displaced continental fragment?: U.S. Geol. Survey Prof. Paper 800-B, p. 211-217.

King, Clarence, 1878, Systematic geology: U.S. Geol. Explor. 40th Parallel Rept., 803 p.

Kozur, Heinz, 1974, Probleme der Triasgliederung und Parallelisierung der germanischen und tethyalen Trias. Teil I: Abgrenzung und Gliederung der Trias: Freiberger Forschungshefte, v. 298, p. 139-197.

Kummel, Bernhard, 1969, Ammonoids of the Late Scythian (Lower Triassic): Harvard Univ. Mus. Comp. Zoology Bull., v. 137, p. 312-542.

MacMillan, J. R., 1971, Time of emplacement and fold history of the upper plate of the Golconda thrust north of 39°N.: Geol. Soc. America Abs. with Programs, v. 3, p. 153-154.

——1972, Late Paleozoic and Mesozoic tectonic events in west central Nevada [Ph. D. Dissert.]: Evanston, Ill., Northwestern Univ., 146 p.

Mosher, L. C., 1968, Triassic conodonts from western North America and Europe and their correlation; Jour. Paleontology, v. 42, p. 895-946.

Muller, S. W., 1949, Sedimentary facies and geologic structures in the Basin and Range province, *in* Longwell, C. R., chm., Sedimentary facies in geologic history [symposium]: Geol. Soc. America Mem. 39, p. 49-54.

Muller, S. W., Ferguson, H. G., and Roberts, R. J., 1951, Geology of the Mount Tobin quadrangle, Nevada: U.S. Geol. Survey Geol. Quad Map GQ-7.

Multer, H. G., and Hoffmeister, J. E., 1968, Subaerial laminated crusts of the Florida Keys: Geol. Soc. America Bull., v. 79, p. 183-192.

Nichols, K. M., 1971, Overlap of the Golconda thrust by Triassic strata, north-central Nevada: Geol. Soc. America Abs. with Programs, v. 3, p. 171.

——1972, Triassic depositional history of China Mountain and vicinity, north-central Nevada [Ph. D. Dissert.]: Stanford, Calif., Stanford Univ., 142 p.

——1974, Coextensive supratidal dolomite and underlying secondary dolomite in the Triassic of north-central Nevada: Jour. Sed. Petrology, v. 44, p. 783-789.

Purdy, E. G., 1963, Recent calcium carbonate facies of the Great Bahama Bank: Jour. Geology, v. 71, p. 334-355, 472-497.

Schweickert, R. A., and Cowan, D. S., 1975, Early Mesozoic tectonic evolution of the western Sierra Nevada, California: Geol. Soc. America Bull., v. 86, p. 1329-1336.

Silberling, N. J., 1956, "Trachyceras zone" in the Upper Triassic of the western United States: Jour. Paleontology, v. 30, p. 1147-1153.

——1959, Pre-Tertiary stratigraphy and Upper Triassic paleontology of the Union District, Shoshone Mountains, Nevada: U.S. Geol. Survey Prof. Paper 322, 67 p.

——1961, Upper Triassic marine molluscs from the Natchez Pass formation in northwestern Nevada: Jour. Paleontology, v. 35, p. 535-542.

——1962, Stratigraphic distribution of Middle Triassic ammonites at Fossil Hill, Humboldt Range, Nevada: Jour. Paleontology, v. 36, p. 153-160.

——1970, Thrust faulting of the Havallah sequence in the Sonoma Range, north-central Nevada: Geol. Soc. America Abs. with Programs, v. 2, p. 143-144.

——1973, Geologic events during Permian-Triassic time along the Pacific margin of the United States, *in* Logan, A., and Hills, L. V., eds., The Permian and Triassic Systems and their mutual boundary: Canadian Soc. Petroleum Geologists Mem. 2, p. 345-362.

——1975, Age relationships of the Golconda thrust fault, Sonoma Range, north-central Nevada: Geol. Soc. America Spec. Paper 163, 28 p.

Silberling, N. J., and Roberts, R. J., 1962, Pre-Tertiary stratigraphy and structure of northwestern Nevada: Geol. Soc. America Spec. Paper 72, 58 p.

Silberling, N. J., and Tozer, E. T., 1968, Biostratigraphic classification of the marine Triassic in North America: Geol. Soc. America Spec. Paper 110, 63 p.

Silberling, N. J., and Wallace, R. E., 1967, Geology of the Imlay quadrangle, Pershing County, Nevada: U.S. Geol. Survey Geol. Quad. Map GQ-666.

——1969, Stratigraphy of the Star Peak Group (Triassic) and overlying lower Mesozoic rocks, Humboldt Range, Nevada: U.S. Geol. Survey Prof. Paper 592, 50 p.

Silver, L. T., and Anderson, T. H., 1974, Possible left-lateral early to middle Mesozoic disruption of the southwestern North American craton margin: Geol. Soc. America Abs. with Programs, v. 6, p. 955-956.

Smith, J. G., McKee, E. H., Tatlock, D. B., and Marvin, R. F., 1971, Mesozoic granitic rocks in northwestern Nevada: a link between the Sierra Nevada and Idaho batholiths: Geol. Soc. America Bull., v. 82, p. 2933-2944.

Speed, R. C., 1971a, Golconda thrust, western Nevada: regional extent: Geol. Soc. America Abs. with Programs, v. 3, p. 199-200.

——1971b, Permo-Triassic continental margin tectonics in western Nevada: Geol. Soc. America Abs. with Programs, v. 3, p. 199.

——1977, Island-arc and other paleogeographic terranes of late Paleozoic age in the western Great Basin, *in* Stewart, J. H., Stevens, C. H., and Fritsche, A. E., eds., Paleozoic paleogeography of the western United States: Soc. Econ. Paleotologists and Mineralogists, Pacific Sec., Pacific Coast Paleogeography Symposium I, p. 349-362.

Stewart, J. H., and McKee, E. H., 1977, Geology and mineral resources of Lander County, Nevada, Pt. 1: Nevada Bur. Mines Bull. (in press).

Tozer, E. T., 1971, Triassic time and ammonoids: Problems and proposals: Canadian Jour. Earth Sci., v. 8, p. 989-1031.

Wallace, R. E., Silberling, N. J., Irwin, W. P., and Tatlock, D. B., 1969a, Geologic map of the Buffalo Mountain quadrangle, Pershing and Churchill Counties, Nevada: U.S. Geol. Survey Geol. Quad. Map GQ-821.

Wallace, R. E., Tatlock, D. B., Silberling, N. J., and Irwin, W. P., 1969b, Geologic map of the Unionville quadrangle, Pershing County, Nevada: U.S. Geol. Survey Quad. Map GQ-820.

Willden, Ronald, 1964, Geology and mineral deposits of Humboldt County, Nevada: Nevada Bur. Mines Bull. 59, 154 p.

Willden, Ronald, and Speed, R. C., 1974, Geology and mineral deposits of Churchill County, Nevada: Nevada Bur. Mines Bull. 83, 95 p.